SEED FROM MADAGASCAR

SEED FROM

MADAGASCAR

By

Duncan Clinch Heyward

illustrations by
Carl Julien

with a new introduction by
Peter A. Coclanis

UNIVERSITY OF SOUTH CAROLINA PRESS
Published in cooperation with the
Institute for Southern Studies and the
South Caroliniana Society of the
University of South Carolina

SOUTHERN CLASSICS SERIES
JOHN G. SPROAT, *General Editor*

King Cotton and His Retainers:
Financing and Marketing the Cotton Crop of the South, 1800–1925
By Harold D. Woodman

The South as a Conscious Minority, 1789–1861:
A Study in Political Thought
By Jesse T. Carpenter

Red Hills and Cotton:
An Upcountry Memory
By Ben Robertson

John C. Calhoun: American Portrait
By Margaret L. Coit

The Southern Country Editor
By Thomas D. Clark

A Woman Rice Planter
By Elizabeth Allston Pringle

Let My People Go:
The Story of the Underground Railroad and the Growth of the Abolition Movement
By Henrietta Buckmaster

Ersatz in the Confederacy:
Shortages and Substitutes on the Southern Homefront
By Mary Elizabeth Massey

Seed from Madagascar
By Duncan Clinch Heyward

Published in Columbia, South Carolina, by the
University of South Carolina Press in cooperation with
the Institute for Southern Studies and
the South Caroliniana Society

Manufactured in the United States of America

Library of Congress Cataloging-in-Publication Data

Heyward, Duncan Clinch, 1864–1943.
    Seed from Madagascar / by Duncan Clinch Heyward ; illustrations by
Carl Julien ; with a new introduction by Peter Coclanis.
        p.   cm. — (Southern classics series)
    "Published in cooperation with the Institute for Southern Studies
and the South Caroliniana Society of the University of South
Carolina."
    Includes index.
    Previously published: Spartanburg, S.C. : Reprint Co., 1972.
    ISBN 0–87249–894–8
    1. Rice—South Carolina. 2. Plantation life—South Carolina.
3. Afro-Americans—South Carolina. 4. Slavery—South Carolina.
5. South Carolina—History. I. University of South Carolina.
Institute for Southern Studies.   II. South Caroliniana Society.
III. Title. IV. Series.
F269.H48   1993
975.3′03—dc20   92–21426

CONTENTS

CONTENTS

GENERAL EDITOR'S PREFACE

T HE Southern Classics Series returns to general circulation books of importance dealing with the history and culture of the American South. Under the sponsorship of the Institute for Southern Studies and the South Caroliniana Society of the University of South Carolina, the series is advised by a board of distinguished scholars, whose members suggest titles and editors of individual volumes to the general editor and help to establish priorities in publication.

Chronological age alone does not determine a title's designation as a Southern Classic. The criteria include, as well, significance in contributing to a broad understanding of the region, timeliness in relation to events and moments of peculiar interest to the American South, usefulness in the classroom, and suitability for inclusion in personal and institutional collections on the region.

*　　*　　*

No economic enterprise did more to shape the society and history of the coastal South, especially before the Civil War, than rice cultivation. And no memoir tells more about rice and the distinctive cultures it spawned, white and black, than Duncan Clinch Heyward's *Seed from Madagascar*. Peter A. Coclanis's introduction to this new edition is not only a thoughtful appreciation of a southern classic, but as well a superb reminder of the critical importance of rice culture in the larger world.

John G. Sproat
General Editor, *Southern Classics Series*

INTRODUCTION

I F history is a graveyard of aristocracies, as Pareto claimed, the South Carolina low country would seem a good place to prove the point.[1] For nowhere was the South's planter aristocracy so powerful nor its reign so long as in this strange and eerie land of silent, still rivers and dark, funereal swamps. In this region a cohesive and confident planter aristocracy held hegemonic power for well over a century, with the area's great rice-planting families the most vivid expressions of the strength of this quasi-hereditary ruling elite. But the time of even this aristocracy eventually passed, and its members succumbed to, merged with, or metamorphosed into new "modernizing" constellations in the late nineteenth and twentieth centuries. It is the cradle-to-grave story of this rice-planting aristocracy that Duncan Clinch Heyward sought to tell in his revealing and some-times poignant 1937 memoir *Seed from Madagascar*.[2] In so doing Heyward—one of the last of the great rice planters—tells us many other important and interesting stories as well.

The aristocracy about which Heyward wrote went back a long way. It seemed older still. Indeed, even in the antebellum period, the low-country aristocracy suggested venera-

bility to some, even superannuation, because of its man-
nered refinement, its soft Adamesque look, its uncalloused
feel.[3]

But for all its ancient usages, country seats, and coats of
arms, the low-country aristocracy, like all others, lived on
invented tradition.[4] The huntsmen and drawing-room phi-
losophes of the 1850s were the grandsons of the parvenus
and nabobs of the 1770s, men whose fathers were—or men
who were themselves—familiar with the back room of the
counting house and the mud of the fields.[5]

The low-country aristocracy was not timeless, then: its
roots lay in the early eighteenth century with the beginnings
of the region's plantation economy. Long before there was
a race season or a St. Cecilia Society or a South Carolina
College or a St. Michael's Church, there was rice and there
were slaves, the twin pillars upon which the low-country
aristocracy was built.[6]

To be sure, there were some wealthy individuals in South
Carolina from the start. It is clear, moreover, that the process of capital accumulation in the low country began well
before the commercial cultivation of rice. Indeed, the ex-
ploitation of slave labor began in the low country before any
agricultural staple counted for much, and was itself a man-
ifestation of this process. Nevertheless, one must take pains
to point out that it was rice, grown by African slaves on
black-water swamps, that established the long-term viability
of the low country, making a local aristocracy possible.

Simply put, prior to the advent of plantation agriculture,
the economic activities pursued in the low country were too
rudimentary to place the region on solid ground. That such

activities were rationally conceived and, given local factor constraints, understandable, does not negate this fact. Furthermore, such activities may have whet, but in no way satisfied, the mercantilist imagination. Obviously, there was money to be made by provisioning Barbados, trading with pirates, and "harvesting" Indian slaves, but such activities were hardly the stuff of a would-be aristocracy. The eager and expectant lot that settled South Carolina hoped for bigger things, as did the Crown: a staple crop as in mercantilist colonies such as Virginia and Jamaica, for example.

After years of frustrating trials with a variety of plants and herbs, the settlers got what they desired, if not deserved: a staple with long-term potential, namely, rice. In seizing upon this commodity in the 1690s, they believed that they had found a way at once to regularize and systematize the accumulation process in the low country, to enrich a sizable proportion of the white population in the area, and to propitiate the Crown. Thus, the seed from Madagascar was sown.[7]

For a long time these expectations bore up well. Rice did in fact discipline the hitherto unruly process of accumulation, many Carolinians did become wealthy on rice and slaves, and the first three Georges were nothing if not satisfied with South Carolina's place in the imperial world. And well they should have been. Of the major mainland colonies, South Carolina had the closest trading relationship with Great Britain, and on the eve of the Revolution the low country was the wealthiest region in British North America by far.[8]

Moreover, both the low country and the aristocracy ger-

minating therein continued to flourish well after the Revolution, with the health of each based primarily on rice and
slaves. Ultimately, however, the low-country economy col-
lapsed and this aristocracy withered, developments which
were related ironically (or perhaps dialectically) to rice and
slaves as well. Let us turn directly, if briefly, to the story of
the low-country rice industry, that we might see what was
reaped from that seed early sown.[9]

First of all, it should be noted that however important
rice was to the low country, it was never king of southern
staples.[10] Tobacco reigned in the eighteenth century and was
succeeded by an obstreperous upstart, cotton, in the nineteenth century. Rice was long part of the royal retinue,
though, and is in fact still an important crop in parts of the
South today.

Charleston wags have long joked that the low country
shares much with China in that the inhabitants of both of
these places worship their ancestors and eat rice. This may
be true, but the Chinese have been doing these things for a
few thousand years longer.

Domestication of the cereal *Oryza sativa* began in Southeast Asia seven millennia ago, whence it spread to China
and other parts of Asia, the Middle East, Africa, and, much
later, Mediterranean Europe. The cereal was transferred to
the Western Hemisphere only in the early modern period,
as part of the so-called Columbian exchange of biogens.

Some rice may have been grown in Spanish Florida in
the sixteenth century, and the English experimented with
the crop in Virginia in the early seventeenth century. It was
not until the last decade of the latter century, however, that

rice became firmly established in the American South, and it became so neither in Florida nor in Virginia but in the youthful English settlement of Carolina.

From the time of initial settlement in 1670, the white colonists in the precociously commercialized Carolina colony searched hard for a viable export commodity. After more than two decades of experiments, failures, and false starts with various and sundry minerals, raw materials, and plant and animal products, they began to experience some success with rice. The precise origins of rice cultivation in the southern part of the colony—Carolina did not split into two separate entities, North Carolina and South Carolina, until 1729—are controversial but relatively unimportant. Whether one believes that rice cultivation initially owed more to Europeans or to Africans ultimately matters little. The cereal was well-known throughout the Old World by the late seventeenth century, and small quantities had already been grown successfully in the New. Whichever "foundation myth" one prefers—and Heyward clearly prefers the one featuring Captain Thurber, the (rice) seed from Madagascar, and all of that—it was not until the mid-1690s that the colony possessed sufficient stocks of labor, capital, and "local knowledge" successfully to begin cultivating, processing, and marketing a staple agricultural commodity such as rice.

For a short period of time, apparently, rice was grown in Carolina without irrigation on dry and relatively high ground in the low country. By the 1720s, production had shifted almost entirely to freshwater swamps in the area, where rudimentary irrigation works could be employed.

Cultivation remained centered in these inland swamps in the low country of South Carolina and, after about 1750, Georgia until the last quarter of the eighteenth century, when the locus of activity shifted again, this time to swampland on, or adjacent to, the area's principal tidal rivers. Indeed, rice production in South Carolina and Georgia, and, to a lesser extent, in the Cape Fear region of North Carolina and parts of northeastern Florida, became increasingly concentrated geographically in the narrow zone on each of this area's major tidal rivers close enough to the coast to be affected significantly by tidal action but far enough inland to run with fresh water. It was along such rivers—six major ones in South Carolina and five in Georgia—that American rice production would be concentrated until the late nineteenth century, when production shifted increasingly to the Old Southwest. Within the South Atlantic region itself, the low country of South Carolina was always the center of production.

Rice cultivation in South Carolina and Georgia was arduous—the crop demanded a great deal of weeding and hoeing—and was characterized by tight labor controls and considerable coercion throughout its history. No area in the entire South, in fact, was so thoroughly dominated by the institution of slavery as the low country under the rice regime and in no area was the role of African Americans and the influence of African-American culture so profound.

To say this is not to suggest, as some have, that African Americans alone were responsible for the technical evolution of the low-country rice industry. If some slaves were from "rice countries" in West Africa and some technology—fanner baskets, for example—was clearly of African origin, much of the technology employed was generic in nature, and, thus, familiar to cereal producers throughout the world. The origins of even the "task system," the distinguishing feature of labor organization in the low country, are open to question. Under this system, which evolved gradually after the mid-eighteenth century, a slave was responsible for completing a specified amount of work daily, a certain number of specified "tasks," as it were, upon the completion of which he or she was free to do what he or she chose. This system most likely grew out of an ongoing process of informal "negotiations" between laborers, "bargaining" for greater autonomy, and managers, hoping to raise productivity and to lessen labor unrest by injecting the incentive of free time into the labor equation. However uncertain the origins of the system, its results are clear: over time, slaves used the "freedom" gained through the task system to work for themselves and/or to "sell" their "free" time to others. In so doing, they were often able to accumulate considerable amounts of personal property, which was, of course, only one of many ironies under slavery.

In any case, with the shift in the early eighteenth century to irrigation, rice production technology became increasingly elaborate and costly. In combination with the geographical limits imposed by nature, such technological considerations helped to create an agricultural complex dominated by a relatively small number of capital-intensive plantations, which utilized sizable numbers of dependent laborers to produce rice and, at times, other staples, for distant, largely foreign, markets.

The main markets for rice produced in the Southeast were never local. Until the late antebellum period, they generally were not even domestic, for most of the crop produced each year was destined for shipment abroad, particularly to the grain markets of northern Europe. In these markets, rice was viewed as a cheap commodity with numerous uses. Rice was sold as a dietary supplement or complement, for example, and as an animal feed. It was used in distilling and, by the mid-nineteenth century, in brewing, and found employment in the starch, paper, and paste industries. Its most common use, however, was as a source of cheap, bulk calories for the poor, and for soldiers, sailors, inmates, and schoolchildren in the absence of, or instead of, more desirable, but often more expensive, foodstuffs.

Prior to the entrance of American rice in European markets, most of the Continent's supply came from the Italian states of Lombardy and Piedmont or from the Levant. By the mid-eighteenth century, though, rice from South Carolina and, later, Georgia, had supplanted other suppliers in the principal European markets and American rice maintained this position until the 1830s, when exports from the United States were surpassed by those from India and Southeast Asia. Small quantities of rice from India had long found their way to Europe, of course, but it was not until the last decade of the eighteenth century that Asia, the Subcontinent in particular, was sufficiently integrated into European trading networks as to allow for the profitable exchange of a bulky, low-unit-value commodity like rice. From that time on, rice from Asia, first primarily from Bengal and, later, from Java, Lower Burma, and Siam as well, poured into European entrepôts such as London, Amsterdam, Rotterdam, and Hamburg.

Given the character of European demand, it is not surprising that Southeast Asia, the lowest-cost supplier in the market, could compete successfully with other supply sources, especially in light of the striking improvements in transoceanic navigation and communications during the nineteenth century. As a result of Southeast Asian penetration of its major markets, American producers shifted their attention in the late antebellum period to the domestic market and to other markets in the Western Hemisphere, most notably Cuba. Despite some success with this strategy, the rice industry of the South Atlantic states was clearly mature well before the Civil War: the rate of growth in output was slowing down, soil fertility was declining, costs (particularly for labor) were rising, exports during the decade of the 1850s were no greater than they had been during the 1790s, and profit possibilities in the industry were diminishing. All of this before the disruption, destruction, and dislocations of four years of civil war.

Until recently, historians believed that the problems of the South Atlantic rice industry began in 1861 and that the industry's demise was a direct outgrowth of the Civil War and emancipation. It is now clear, however, that its problems were both structural and long-term in nature, having as much to do with the expansion and elaboration of capitalism and with shifts in international comparative advantage, as with federal occupation, wartime destruction of production facilities, and postwar shortages of capital and changes in labor relations.

To be sure, the short-term factors disrupted and impeded the South Atlantic rice industry. Production in the four South Atlantic states of North Carolina, South Carolina, Georgia, and Florida fell from an all-time high of 179.4 million pounds of clean rice in 1859 (95.9 percent of the U.S. total) to only 57 million pounds in 1869. But production in the area rose by nearly 48 percent between 1869 and 1879, and even as late as 1899 almost 69 million pounds of clean rice were produced in the South Atlantic region. By that time, however, Southeast Asian competition had not only knocked American rice out of Europe, but had penetrated the domestic market as well. The United States, in fact, was a major importer of rice for a half century after the Civil War.

One important long-term result of such competition was the gradual geographic migration of the rice industry to the Old Southwest, that is, to Louisiana, Texas, and Arkansas. Here, highly mechanized production technology was employed, particularly after the so-called rice revolution of the mid-1880s, which raised productivity and minimized the problems posed by scarce and/or unruly labor. Though rice *could* still be grown in the South Atlantic region at the turn of the century, highly mechanized production technology could not be, or at least was not, introduced on the reconstituted plantations, tenant plots, and scratch freeholdings in the area. Consequently, production, generally speaking, no longer meant profits, and the low country of South Carolina and Georgia lapsed into generations of stagnation and decline. In the last analysis, however, the evolution of the rice industry owed as much to European imperialism and to development in Calcutta, Batavia, and Rangoon, as to more familiar events closer to home.

The low-country aristocracy—at least *a* low-country aristocracy—failed to survive these changes intact. The region, after all, had been built by and for rice. Other activities—the plantation production of indigo and long- and short-staple cotton, the collection and processing of forest products, manufacturing, and the provision of commercial services—were always of secondary importance, complementing but never challenging the rice regime, whose power was well-nigh absolute.

With the collapse of rice, the aristocracy that rice built was forced at once to confront an economic crisis, a moral crisis, and a crisis of legitimation. Not surprisingly, it splintered. The aristocracy that was so bold, young, and harmonious during the Revolutionary era, that was older but no less unified and audacious in 1860 and 1861, now seemed shaken, uncertain whether to resist, to change, or to run. That none of these options seemed particularly attractive was perhaps most unsettling of all. For none realistically could prevent the disappearance of a class and a way of life that its fearsome and awful ancestors had begun.[11] Hence, the wistfulness, the sense of failure, the existential sadness that mark insurance broker Duncan Clinch Heyward's oddly moving book. For someone born into one of South Carolina's first families, someone who had once been a great rice planter and twice been governor, selling group policies—Heyward's insurance specialty in the 1930s— could not have salved a heart.

It is not easy from our vantage point today to define precisely, much less to understand completely, the aristocracy that gave way in the low country at the turn of the century. Membership was not contingent upon formal title nor, strictly speaking, was status hereditary. There was movement into and out of this aristocracy and no single factor—not wealth or family or character or achievement—guaranteed inclusion.[12]

What one had to possess, it seems, was commanding authority, however attained.[13] If this usually came from lineage and wealth—in the low country, particularly wealth in slaves—it could originate from other sources as well, including the power and cast of one's mind, as with David Ramsay, John Bachman, Bartholomew Carroll, James L. Petigru, and William Gilmore Simms. Generally speaking, though, the low-country aristocracy during the rice regime comprised wealthy planters of good stock with many slaves. If merchants, lawyers, scholars, and theologians were part of this cultural production, they never occupied center stage.[14]

Commanding authority, however, was not all that distinguished the low-country aristocracy. Small numbers and intricate—some would say involuted—affective and kinship ties marked it as well. That this aristocracy was small and close is not surprising. The low country was overwhelmingly rural, the white population as a whole in the area was never large, and whites were a minority—in some places, a tiny minority—in the low country by the early eighteenth century.

Although precise estimates of the size of the aristocracy are difficult to come by, and, in any case, would be misleading if they were employed, what we are talking about is a number of households ranging in the hundreds rather than in the thousands. Over 60 percent of the 440 South Carolina planters with more than one hundred slaves in 1860 lived in the low country, for example. Most of these planters were members of the low-country aristocracy, as were some slaveholders, obviously, with fewer than one hundred slaves. The point is that even as late as 1860 we are talking about a relatively small number of people.[15]

With so few people involved, it is also not surprising that this aristocracy was closely knit. There was disagreement and there were feuds, to be sure, but regular, indeed, incessant intermingling and intermarriage within the ranks of the aristocracy helped to create a strong sense of common purpose, one similar in some ways to the medieval concept of *noblesse,* or the community of the well-born.[16]

Small numbers and close interrelationships along with the stabilizing, at times even immobilizing, tendencies associated with rice cultivation, helped to add another distinguishing characteristic to the low-country aristocracy: continuity of membership. Despite inward and outward mobility, many of the same families were represented for long, long periods of time, which by the late-antebellum period gave this aristocracy something of a *rentier* tinge. Indeed, among those planters with more than one hundred slaves in 1860 we find many descendants of families that were already part of the planter elite in the Revolutionary era. Thus, we find the names Ball, Barnwell, Blake, Bull, Cordes, Deas, Drayton, DuBose, Elliott, Fripp, Gaillard, Gourding, Horry, Huger, Hume, Izard, Ladson, Lowndes, Mani-

gault, Matthewes, Middleton, Moore, Moultrie, Pinck-
ney, Porcher, Prioleau, Ravenel, Rutledge, Screven, Simons, Trapier, VanderHorst, Waring. And, of course, the
name Heyward too.[17]

Although the exact date of his arrival is unknown, Daniel
Heyward, the first Heyward in South Carolina, was living
in the colony by 1684, having arrived perhaps as early as
1672. In any case, he was there well before the age of rice
began.[18] Members of the Heyward family quickly became
active in local politics, but the family did not become identified with the production of rice for several generations.
According to the author of *Seed from Madagascar,* Daniel
Heyward (1720–1777), great-grandson of his namesake,
was the first Heyward to plant rice in South Carolina, when
he began cultivation in the early 1740s of a large tract of
land on Hazzard's Creek in St. Helena's Parish. He proved
extremely successful in his rice-planting activities, and in
1771 the *South-Carolina Gazette* reported that he was "the
greatest Planter in this Province." He may well have
been—he was said to have owned about one thousand slaves
at the time of his death—but his successes, such as they
were, paled in comparison to those of his son Nathaniel,
who was to become the largest planter in the entire history
of the Slave South.[19]

Nathaniel Heyward, the son of Daniel Heyward and Jane
Elizabeth Gignilliat (the Gignilliats were another prominent low-country family) was born in 1766. As a boy during the American Revolution he served on the side of the
Patriots, and, upon reaching his majority, inherited several

INTRODUCTION

small tracts of land from his father. Over time he trans-
formed his relatively modest inheritance into an estate of
fabulous size. At one point he is said to have owned over
2,500 slaves, and in 1849, two years before his death, the
federal census reported that he owned 35,000 acres of land
in the low country (mainly in the Colleton and Beaufort
districts), including enough rice land to produce 16.7 million pounds of (rough) rice in that year. In 1849, moreover, his 20-odd low-country plantations were home to 100
working oxen, 600 "milch" cows, 2,000 other cattle, 1,000
sheep, 1,000 swine, and 80 horses, mules, and asses, which
offers further evidence, if any were needed, of his operations' scale, scope, and size. When he died in 1851, he still
owned roughly 2,000 slaves, and a total estate, slaves included, valued at just over $2 million. In monetary terms,
then, Daniel Heyward's son—Duncan Clinch Heyward's
great-grandfather—had done very well indeed.[20]

Nathaniel Heyward was not nearly so fortunate in his
family life. His wife, Harriet (Henrietta) Manigault,
whom he had married in 1788, died in 1827, leaving Heyward a widower for the last twenty-four years of his long
life. Moreover, only four of their nine children—two sons
and two daughters—survived him. Upon his death, his vast
estate was divided among his surviving children and his
many grandchildren, which left Arthur, his youngest son,
with a magnificent rice plantation, called the Bluff, on the
Combahee River, and Charles—Duncan Clinch Heyward's
grandfather—with four large rice plantations on the same
body of water.[21]

Charles Heyward was born in Charleston in 1802 and

spent several years at Princeton, though he never graduated. After returning to South Carolina, he helped manage his father's many plantation enterprises, taking up residence at Rose Hill Plantation, which adjoined the Bluff, Nathaniel Heyward's homeplace on the Combahee. At the age of twenty-one, Charles Heyward married Emma Barnwell of the famous Beaufort family of Barnwells, and the marriage produced three sons and two daughters (another daughter died in infancy) before Emma's premature death in 1835 at the age of twenty-nine. Like his father, Charles Heyward never remarried, spending the last thirty-odd years of his life a widower.[22]

Also like his father, Heyward was a serious and careful rice planter, whose operations on the Combahee were profitable even after the low-country rice industry had begun its long decline. The four plantations he inherited in 1851 from his father—Amsterdam, Lewisburg, Pleasant Hill, and Rose Hill—he operated successfully well into the Civil War. According to the federal census of 1860, Charles Heyward owned 471 slaves in that year, though Duncan Clinch Heyward later claimed that his grandfather owned 529 slaves in the late 1850s and 495 at the time of emancipation. Whichever number one prefers, it is clear that Charles Heyward was a very, very large planter.[23]

After Union forces began their occupation of the low country, Heyward and most of his slaves quit the Combahee region for an inland plantation near Columbia owned by one of his sons. He never again saw his beloved Rose Hill, for he died in exile in March 1866, less than a year after Appomattox.[24]

The inland plantation upon which Heyward had sought refuge was owned by his eldest son, Edward Barnwell Heyward. Barnwell Heyward was born in Beaufort, his mother's hometown, in 1826, and graduated South Carolina College in 1845. After a European tour, he returned to the Palmetto State, and in 1850 he married Luzy Izard of Columbia, in which city the couple took up residence. Hopeful of improving his wife's fragile health, Barnwell Heyward later purchased and moved his family to a plantation known as Goodwill, located about twenty-two miles outside of Columbia on the Wateree River. Here Heyward and his slaves planted cotton and corn, as well as a little rice, which must have comforted Charles Heyward to a degree during his war-imposed exile from the low country.[25]

This move to the country notwithstanding, Lucy Izard Heyward died at an early age, leaving her husband to raise a son, Walter Izard Heyward, who had been born in 1851; the couple's two other children, both girls, had predeceased their mother, both having died in infancy. In 1863 Barnwell Heyward broke with family tradition and remarried. His second wife, Catherine Maria Clinch, not surprisingly perhaps, came from a rice-planting family, though her rice-planting forebears (on her mother's side) were from Georgia rather than South Carolina.[26]

During the war, Barnwell Heyward served as a lieutenant of engineers in the Confederate Army, and, with his father's death in 1866, he inherited two Combahee River rice plantations, Lewisburg and Amsterdam. In the chaotic period immediately after the war, Heyward succeeded, after strenuous effort, in reestablishing production on these plan-

tations. Neither he nor his wife lived long enough, however, to see what his efforts ultimately wrought: Catherine
Heyward died in 1870 in her thirty-eighth year, and Barnwell Heyward, aged forty-four, died shortly afterward, in
1871. They left behind two young sons, Bayard Clinch
Heyward, born in 1867, and our protagonist, Duncan
Clinch Heyward, a war baby, born on June 24, 1864.[27]

Within a year of "Clinch" Heyward's birth the war
ended, and in the fall of 1866 Barnwell Heyward and his
family moved to the low country to lay claim to the Heyward properties on the Combahee. Given the unsettled conditions in the countryside, the family took up residence in
Charleston, but young Clinch spent a good deal of time at
Lewisburg and Amsterdam, which were still formidable
plantations.[28] If the size and scale of Barnwell Heyward's
holdings could not compare with those of earlier generations
of Heywards, it should be noted nonetheless that, taken together, Lewisburg and Amsterdam contained roughly 800
acres of rice land and 1,500 acres of so-called highlands,
not too shabby an estate, provided sufficient labor could be
found.[29] A sizable loan from an Ohio capitalist helped secure an adequate supply of labor, and the plantations were
back in business a few years after the war.[30]

But for the death of his parents, Clinch—a strong, ath-
letic boy with a bent for the outdoors—would likely have
spent his youth in the same way as did many of his forebears
on the Combahee: learning the mysteries of the planter's art.
With his father's death, however, young Heyward was
taken in by Eliza Bayard Clinch, his maternal grand-

mother, who at the time wintered in Charleston, but summered in mountainous Habersham County, Georgia.[31] Both the mysteries of planting and the mysterious Combahee would have to wait for years to come.

Clinch Heyward was educated at private schools in Charleston before moving north to attend Cheltenham Academy, a private boarding school in Shoemakertown, Pennsylvania, near Philadelphia.[32] Despite or perhaps because of the curriculum at Cheltenham, Heyward developed an interest in belles lettres, including poetry, and several awkward, indeed, awful poems written by Heyward during his teenage years still survive. "Sunset at Lamont," written when Heyward was sixteen, offers a case in point.

Sunset at Lamont

The western sky is tinged with red,
As sinks the sun o'er Yonah's head.
The mountains all with bright color glow,
And cast their shades in the vale below.

So day by day that Sun's gone down,
And lit the landscape all around.
But never a brighter sky I ween,
Than when by me the last time seen.

Now silence broods o'er vale and hill,
No sound is heard but the whippowill.
Soon, soon, Lamont again I'll see,
And Oh! how happy at home I'll be.[33]

However maudlin those verses, "Going to War," a poem Heyward wrote two years later, represents an even more highly developed state of mawkish sentimentality. One verse of this gem should suffice.

Going to War

Farewell, dearest wife!
The time has come to part.
To stay, I'd almost give my life.
To go, it breaks my heart.[34]

Fortunately for both Heyward and American letters, the youth's infatuation with poetry seems to have ended shortly thereafter.

After spending three years at Cheltenham, Heyward headed south to attend Washington and Lee University in Lexington, Virginia, where he studied for three years before returning to South Carolina in 1885.[35] Once back in South Carolina, Heyward began preparing himself to take up his share of Amsterdam and Lewisburg, the two Combahee River rice plantations he had inherited from his father. When Barnwell Heyward died all three of his sons were still minors—Walter was approaching adulthood—and his rice plantations were entrusted to the care of Colonel Allen Cadwallader Izard, the brother of his first wife, who managed them for the next eighteen years. Izard, a graduate of the Naval Academy and an ex-Confederate officer, was a fine planter, according to Clinch Heyward, and "an excellent manager of rice-field Negroes." During his tenure, he was successful enough in his planting activities both to repay the debt Barnwell Heyward had taken on in reestablishing production after the war and to build up the principal in the Heyward estate.[36]

Moreover, upon Clinch Heyward's return to the Combahee, Izard, along with long-time overseer Squire Jones, taught the young man at long last the planter's art. By his own admission, Heyward "knew nothing about rice planting" when he left Washington and Lee. If he had "always believed that [he] was destined to plant the fields which [his] family had planted," beliefs don't make a crop. Without the guidance and instruction of Izard and Jones—and, later, another overseer, William Jaycocks, as well—Heyward probably would have remained an armchair agriculturalist or a planter *manqué* at best.[37]

"I am trying to learn all I can about planting, and am watching things as closely as possible," he wrote his uncle Houstoun Clinch in April 1887. Heyward apparently learned much and watched well: later in the year he made the first of his twenty-six crops.[38]

If certain details remain unclear about his specific agricultural practices, surviving sources—letters and plantation account books primarily, but also *Seed from Madagascar*—indicate clearly that Heyward was a far better planter than he was a poet. Over the quarter century between 1887 and 1912, Heyward generally planted about 500 acres of Combahee River rice land annually, and generally achieved yields of 36 to 40 bushels per acre. Despite a variety of long-run and short-term problems ranging from low prices and foreign competition to recalcitrant labor to hurricanes, bank breaks, and floods, Heyward's receipts from rice easily covered his direct planting expenses in most years.[39]

Nonetheless, he felt—or at least claimed—that he had betrayed both his family and his class when he abandoned rice planting in 1913. In reality, he had actually acquitted himself quite well in his planting operations, particularly in comparison to others of his ilk. Furthermore, given his ancestors' resourcefulness and ambition, they likely would have lauded rather than reproached him for the nimbleness he displayed in leaving agriculture (he sold his land to a Du Pont for use in part as a hunting preserve) and entering the modern bourgeois world. For the line separating the Big House from the counting house was a rather casual one in the low country all along.[40]

Unlike a nineteenth-century aristocrat such as Tomasi di Lampedusa's fictional Don Fabrizio, a sagacious, premodern Sicilian prince who slowly went down with his world, Clinch Heyward moved purposively and decisively when he had to and always had an eye on the main chance. Despite his family's long-time identification with the cultivation of rice, Heyward himself, protests notwithstanding, seems to have viewed rice-planting as a way to make a living rather than as a way of life. Thus, he was easily able, when the time came, to break the bonds tying him and his family to this most intractable and inertial hydraulic crop.[41]

From the time Clinch Heyward commenced his "career" as a rice planter, he was well aware of, and engaged in, a wider world. He and his family—he had married Mary Elizabeth Campbell, the daughter of a Virginia planter, in February 1886, and their union ultimately produced four children—chose to live in the Colleton County town of Walterboro, roughly a dozen miles from the Combahee, rather than on one of the family plantations. While living in Walterboro, the county seat, Heyward involved himself in a variety of clubs and organizations ranging from the fraternal—the Knights of Pythias and the Masons, for example—to the parliamentary: he was for a time captain of the Combahee Mounted Riflemen, which had been formed to keep the area's black majority in line.[42] During the same period, he found time to dabble in real estate development in the North—in a subdivision of Sioux City, Iowa, of all places—and, of course, to keep track of his planting operations, which were managed on the day-to-day level by Squire Jones until January 1889, and after that date, by William Jaycocks, Jones's son-in-law.[43]

Moreover, Heyward increasingly found time for another activity: politics. While he had been elected captain of the Combahee Mounted Riflemen and chosen grand chancellor of the Knights of Pythias of South Carolina for 1897–1898, Heyward had never stood for any public political office until 1902, when, incredibly, he was elected governor of the Palmetto State.

In the hothouse atmosphere of South Carolina during the Tillman era, the fact that Heyward was a political amateur was not necessarily to his disadvantage. He was untainted by the roily politics of the day—one of his opponents in the gubernatorial race, James H. Tillman, would murder the editor of the state's most influential newspaper shortly after the 1902 elections—and no one knew exactly where Heyward stood on the major issues. Responding in 1903 to queries about why he had run for governor, Heyward stated: "I ran because my friend wished me to make the race and because I thought I could be elected." A year and a half later he elaborated further: "It was my ambition to be governor of South Carolina and to do something for the people of my state."[44]

Beneath the superficial banality of these statements are certain points worthy of note. Heyward's friends—and his family background and clubable nature assured that he had many around the state—did, in fact, push his candidacy, in the belief that his personal integrity, genial demeanor, and "conservative modernizing" worldview would prove beneficial in strife-ridden South Carolina. Moreover, Heyward did indeed wish to do something for—or perhaps about—the people of the state. In effect, he wished to ease their transition to modernity, to get them to accept at least the veneer of bourgeois order, and to help them accommodate to, even as he attempted to moderate, the process of industrialization in the state. Such wishes, put differently, were those of a planter-businessman acting as Bourbon proto-progressive.[45]

After leading a five-man field in the first Democratic primary in June 1902, Heyward defeated Congressman Jasper Talbert of Parksville in a two-man run-off in August, winning almost 56 percent of the vote. In 1904 Heyward won reelection without opposition.[46]

As governor he pursued his ambitions with mixed results. He advocated and pushed for a variety of programs relating to economic development, good government, state fiscal reform, compulsory education, child labor reform, and greater law and order, the last, most notably, through reform of the state's controversial liquor dispensary system and through efforts to reduce, or at least to rationalize, lynching. If little actually was accomplished during his tenure—in a shocking but nonetheless telling incident, for example, a black man was lynched near Greenwood just after the governor had spoken to the lynch mob in hopes of heading off the action—Heyward's programs in many cases set the tone for later "progressive" governors such as Richard I. Manning and Robert A. Cooper.[47]

Whether he was disillusioned with electoral politics or felt his political tasks complete, Heyward never again ran for public office after 1904. Upon completing his second gubernatorial term in 1907, he moved his family to Charleston, but he retained close ties to Columbia in order to pursue the types of business opportunities that often open up for ex-governors. He served terms as president of both the Standard Warehouse Company and the Columbia Savings Bank and Trust, for example, and was for a time a director of the National Loan and Exchange Bank of Columbia.[48]

Once out of politics, Heyward gave his planting operations a final chance, but in 1913, in the aftermath of devastating storms in 1910 and 1911, he finally sold off his properties on the Combahee—Lewisburg, Amsterdam, and another plantation, Rotterdam, he had only recently purchased—ending his "career" as a rice planter. In that same year President Woodrow Wilson appointed him collector of internal revenue for the district of South Carolina, an appointment he held through the election of 1920.

Heyward spent the remainder of his working life in Columbia, engaged for the most part in the brokerage and insurance industries. He was president at one point of his own stock brokerage, D. C. Heyward & Company, and later served as special agent for group insurance for a number of insurers, including Aetna and the Protective Life Insurance Company. At the time of his death in Columbia on January 23, 1943, he lived on Kilbourne Road in the affluent Heathwood section of the city.[49]

Around the time Heyward sold his properties on the Combahee, thereby forfeiting any residual claim to planter status, he began to think about writing a history of his family, the low-country aristocracy, and the region's rice industry.[50] His early poetry notwithstanding, Heyward had long been interested in writing, and an audience seemed to exist for the type of book he had in mind; the warm reception given to Elizabeth Allston Pringle's *A Woman Rice Planter*, published by Macmillan in 1913, suggested as much.[51] Moreover, such a book project might help to assuage his sense of guilt—or possibly shame—over selling the family lands, and provide a creative outlet for a sensitive, middle-aged man locked into a series of prestigious, but humdrum, jobs in an increasingly middle-class world. However lucrative such positions, none apparently—not even corporate presidencies or bank directorships—was capable of stirring Heyward's imagination. On the other hand, the story of rice and of the grandees who shaped its history in the low country seemed capable of doing so, and Heyward turned, atavistically perhaps, to his planting past for both subject material and creative sustenance, even as he looked elsewhere for his material support.[52]

Thus, by 1920 Heyward was well into research on his family's history, research which he broadened and deepened in the ensuing decade. In 1927 he "came out" as a literary man, publishing in South Carolina's leading newspaper, *The State* [Columbia], a series of "droll" sketches relating to "incidents of his public career."[53] Heyward continued to work on his book in the early 1930s, and completed a draft of the entire manuscript by 1935. With the help of an agent of sorts, J. T. Gittman, owner of Gittman's Book Shop in Columbia, Heyward shopped the manuscript at a number of commercial houses, but after rejections from both Macmillan and Dutton, he turned to the University of North Carolina Press, which was then making a name for itself under the inspired leadership of William T. Couch.[54]

UNC Press was enthusiastic about Heyward's manuscript from the start and, upon receiving two favorable reader's reports, one of which, interestingly enough, was from J. G. deRoulhac Hamilton—Professor of History at UNC, founder of the university's Southern Historical Collection, and a relative of the author—Heyward was offered a contract.[55] After a number of small problems had been ironed out—questions relating primarily to the publication schedule, the use of illustrations, and the book's title—*Seed from Madagascar* (a title Couch disapproved of initially because of its vagueness) was published in October 1937.[56]

The handsomely produced book received good reviews, including one from the *New York Times*, and sold pretty well, though perhaps not quite so well as was hoped originally.[57] *Seed from Madagascar*'s initial press run was for twenty-five hundred books, but Couch at one point was hopeful that as many as four or five thousand copies might be sold. In his reader's report, Hamilton sounded an optimistic note as well, arguing that the press could expect strong sales because, as he put it, "Governor Heyward is the most popular man in South Carolina and people will buy his book." Indeed, according to Hamilton, Columbia bookseller J. T. Gittman believed that he alone could sell a thousand copies.[58]

As it turned out, *Seed from Madagascar* did, in fact, eventually exhaust its initial press run, but it took a long, long time. Of the 2,505 copies actually printed, 2,372 had been sold, and another 99 distributed gratis, by June 30, 1948.[59] The remainder of the run was gone sometime before February 1, 1961, when the press informed Katherine B. Heyward, the late author's daughter, that the book was now out of print, but that demand was insufficient to justify a second printing. The press softened the blow somewhat by pointing out that Xerox or microfilm copies of *Seed from Madagascar* would be made available "upon demand" by University Microfilms of Ann Arbor, Michigan. Such copies, however, would be "considerably more expensive than . . . the printed edition."[60] By the early 1960s, then, *Seed from Madagascar*, like its author, like the low-country aristocracy, and like the South Atlantic rice industry, seemed a thing of the past.

But the book did not die, happily, for *Seed from Madagascar* remains an important source of information and insights about a range of topics in southern history. If the book is primarily concerned with the Heyward family and with the low-country rice industry, it has much to tell us as well about black culture and about race relations and about racial attitudes and conventions among the white gentry and bourgeoisie of the South in the late nineteenth and early twentieth century. Moreover, in some ways, *Seed from Madagascar* reads rather like—or at least can be read as— ethnography, and, as such, is full of cultural grist for the postmodern mill.

A few words about matters of form and structure before we go any further, for such concerns are in this case quite revealing. *Seed from Madagascar* is modest in length—248 pages of actual text, no notes, and no bibliography—and is divided into twenty-eight short chapters. Although the author employs a central narrative conceit—the linked trage-dies of the Heyward family and the low-country rice indus-try—the plot is broken occasionally for analytical *discursi* on a variety of topics ranging from the hunting practices of the southern aristocracy to the origins of Gullah speech. Still, the book for the most part is carried on the backs of the Heywards and on those who labored in their fields.

One should note, too, that *Seed from Madagascar* is en-riched, but complicated, by photographic illustrations by noted South Carolina photographer Carl Julien.[61] The pho-tographs and their captions at times subvert, wittingly or unwittingly, the narrative Heyward so carefully constructs. For how are we to take a book which includes a photograph of a black woman and six children on the porch of a wooden shack, the photo captioned "A Study in Local Color"? A book with a shot from the rear of three sad-looking black boys sitting interspaced on a pier, legs dangling over the edge, this photo captioned "Born to be Fishermen"? Or a book with a photo of a black boy in tattered clothes shooting marbles, this one captioned "Always Ready for a Game"? As tragedy? Perhaps, but other nomenclature comes to mind as well.

Nonetheless, *Seed from Madagascar* ultimately derives importance from its substance. Simply put, the book offers one of the best views we have of the operations of a single rice-planting family over time and is, in addition, one of our better sources of information on the technical aspects of rice cultivation in the South Atlantic region. Moreover, Heyward's analysis of the low-country rice industry still seems fresh in many ways. To cite but two examples, he correctly identified the central dilemma nineteenth-century rice planters faced—declining profit possibilities in the low country even in the antebellum period, but few viable alternatives to rice—and he was sufficiently acute to acknowledge, even if he failed to emphasize, the role of Southeast Asian competition in bringing about the low-country rice industry's demise. On Heywards and on rice, then, *Seed from Madagascar* is extremely good, but what about on matters relating to the third panel of this low-country triptych: that is, where do blacks fit in?

Here, again, the book is of demonstrable importance, both because it includes suggestive material on and keen insights into black behavior and ideology, and, alas, because of the questions left unasked, the limitations of Heyward's worldview. Like many low-country whites of his class, Heyward was personally comfortable with blacks, knowledgeable about their culture, and, within the narrow confines of white racialist ideology at the time, relatively well disposed toward them. All of this comes through in *Seed from Madagascar*, which at times seems as much about race as about rice.

Thus, one finds individual chapters on "the Gullah Negro," on field hands and on house servants, and on certain memorable slaves. One finds detailed discussions of black demographic patterns, foodways and housing, and of the nature of childhood under slavery. Topics of great interest today, such as the incidence of property-holding among slaves and questions relating to the ethnicity of slaves, are addressed as well, at times with considerable intelligence and sutlety.[62]

Indeed, both in spite of and because of expressions like "the old darkey" this, and "many a little darky" that, and statements about blacks and watermelons, and about "wild African savages," *Seed from Madagascar* tells us much about blacks and whites alike.[63] One learns about the "Beaufort Manner" adopted by low-country blacks after emancipation, and about "ole Stephney"—hunger—and about why "drivers" became known as "foremen" after the war. One learns that in Heyward's view blacks knew whites as well as, or better than, whites knew blacks, and that blacks pretended "to misunderstand what was said to them when it suited their purpose to do so."[64]

One learns other things too, things not necessarily foreseen by Heyward when writing the book. The indignity of racism is unintentionally underscored, for example, when Heyward relates what he thinks is an amusing tale about a black engineer in a steam-powered threshing mill, who was foolish and audacious enough to believe that he had something to do with the mill's success. And the dignity and indomitability of the human spirit is captured, inadvertently, in another Heywardian divertissement, this one about slaves and their love of shoes: "I have heard that once on a very cold morning, when the ground was frozen, a Negro was seen walking along a road barefooted and carrying a new pair of shoes in his hand. Asked why he did not wear his shoes, he replied, 'Well, oonuh see dese duh me shoesh; me feet dem blonxs tuh Maussuh.' "[65]

But, after reading *Seed from Madagascar*, one begins also to understand—and this is the book's great achievement—what it was about the Heywards, about rice, and about the plantation regime that in 1913, upon word that Clinch Heyward had quit the Combahee, led an ex-slave to weep, to wring her hands, and to scream in agony over and over again: "Oh, me Gawd! Me Maussah 'e ent hab no lan'."[66] For Clinch Heyward, clearly, was a decent man and many of his people were decent as well. If they were also slaveholders and racists, there was plenty of blame to go around.

Given the age in which he lived, it is easy for this writer to believe that Clinch Heyward "was the most popular man" in South Carolina, and even understandable why a mill worker in Honea Path, according to August Kohn, had Heyward's picture on the mantle.[67] I think I can even understand why his sale of land to a Du Pont could have evoked the response it did from an old black woman, an ex-slave at that. For the ex-governor's humanity and broad sympathies come through over and over again in his book.

In the last analysis, Clinch Heyward was a decent man in an indecent world. By the time he wrote his book, however, he was no longer an aristocrat. He was a *bourgeois gentilhomme*, to be sure, but one who had to write a little insurance too. Deep down, this was his complaint, and that is what makes *Seed from Madagascar* an oft powerful lament, but not a tragedy true.[68]

Peter A. Coclanis

Notes

The author would like to thank Lewis Bateman, David Carlton, Lacy Ford, Stuart Leibiger, David Moltke-Hensen, and Bryant Simon for their help with this essay.

1. See Vilfredo Pareto, *The Mind and Society: A Treatise of General Sociology*, trans. Andrew Bongiorno and Arthur Livingston, 4 vols. bound as 2 (New York: Dover Publications, 1963), 3: 1431. The Dover edition is a reprint of the first English translation of Pareto's *Trattato di Sociologia generale*. This translation, published by Harcourt, Brace and Company in 1935, was based on the second edition of the *Trattato* (1923).

2. *Seed from Madagascar* was first published in Chapel Hill by the University of North Carolina Press.

3. This look and feel were best expressed, of course, in antebellum Charleston. See George C. Rogers, Jr., *Charleston in the Age of the Pinckneys* (1969; rpt. Columbia: University of South Carolina Press, 1980), pp. 141–66 especially; Kenneth Severens, *Charleston: Antebellum Architecture and Civic Destiny* (Knoxville: University of Tennessee Press, 1988).

4. On the concept of invented tradition, see Eric Hobsbawm, "Introduction: Inventing Traditions," in Hobsbawm and Terence Ranger, eds., *The Invention of Tradition* (Cambridge: Cambridge University Press, 1983), pp. 1–14.

5. For the best view of intellectual life in the low country in the antebellum period, see Michael O'Brien and David Moltke-Hansen, eds., *Intellectual Life in Antebellum Charleston* (Knoxville: University of Tennessee Press, 1986).

6. On the creation of the low-country aristocracy, see Richard Waterhouse, *A New World Gentry: The Making of a Merchant and Planter Class in South Carolina, 1670–1770* (New York: Garland, 1989); Peter A. Coclanis, *The Shadow of a Dream: Economic Life and Death in the South Carolina Low Country, 1670–1920* (New York: Oxford University Press, 1989), pp. 48–110 especially. On the cultural productions of this aristocracy, also see Frederick P. Bowes, *The Culture of Early Charleston* (Chapel Hill: University of North Carolina Press, 1942); Rogers, *Charleston in the Age of the Pinckneys*, pp. 55–115 especially.

7. See, for example, Coclanis, *The Shadow of a Dream*, pp. 48–110; Converse D. Clowse, *Economic Beginnings in Colonial South Carolina 1670–1730* (Columbia: University of South Carolina Press, 1971).

8. Coclanis, *The Shadow of a Dream*, pp. 48–110.

9. Coclanis, *The Shadow of a Dream*, pp. 111–58.

10. The following discussion on the American rice industry draws heavily from Coclanis, *The Shadow of a Dream*, pp. 133–37; Coclanis, "Rice," in Richard N. Current, ed., *Encyclopedia of the Confederacy* (New York: Simon and Schuster, 1993).

11. Coclanis, *The Shadow of a Dream*, pp. 137–60 especially. See also Coclanis, "Entrepreneurship and the Economic History of the American South: The Case of Charleston and the South Carolina Low Country," in Stanley C. Hollander and Terence Nevett, eds., *Marketing in the Long Run* (East Lansing: Department of Marketing and Transportation Administration, Michigan State University, 1985), pp. 210–19.

12. See Chalmers G. Davidson, *The Last Foray: The South Carolina Planters of 1860: A Sociological Study* (Columbia: University of South Carolina Press, 1971), pp. 1–17 and passim. One should note that the Heywards were one of the few British American families holding a legitimate Patent for Arms. See James B. Heyward, comp., "The Heyward Family of South Carolina," *South Carolina Historical Magazine* 59 (July 1958): 143–58 and (October 1958): 206–23. The information pertaining to the Patent for Arms appears on pages 143 and 144 (footnote 2). It is interesting to note that even W. J. Cash acknowledged the existence of an aristocracy in the South Carolina low country. See Cash, *The Mind of the South* (New York: Alfred A. Knopf, 1941), pp. 4–5.

13. See Jonathan Powis, *Aristocracy* (Oxford: Basil Blackwell, 1984), pp. 8, 43–62 especially. The political-science literature on elites is useful to those interested in studying aristocratic authority. See, for example, T. B. Bottomore, *Elites and Society* (London C. A. Watts, 1964); Geraint Parry, *Political Elites* (London: George Allen & Unwin, 1969).

14. See Davidson, *The Last Foray*, pp. 1–17; David Moltke-Hansen, "The Expansion of Intellectual Life: A Prospectus," in O'Brien and Moltke-Hansen, eds., *Intellectual Life in Antebellum Charleston*, pp. 3–44.

15. On the demographic and economic character of the low country, see Coclanis, *The Shadow of a Dream*, pp. 48–158. The figure in the text pertaining to the residency status of South Carolina's largest planters was derived from biographical material in Davidson, *The Last Foray*. See pp. 170–276 especially.

16. On the concept *noblesse*, see, for example, Powis, *Aristocracy*, p. 8.

17. Davidson, *The Last Foray*, pp. 16–17. For useful information in this regard, see also Mark D. Kaplanoff, "Making the South Solid: Politics and

the Structure of Society in South Carolina, 1790–1815" (Ph.D. dissertation, Cambridge University, 1979).

18. See Heyward, *Seed from Madagascar*, p. 46; Heyward, "The Heyward Family of South Carolina," pp. 144–45; Agnes Leland Baldwin, *First Settlers of South Carolina 1670–1700* (Easley, S.C.: Southern Historical Press, 1985), p. 118. Daniel Heyward, one should note, came to Carolina from the County of Derby, England.

19. Heyward, *Seed from Madagascar*, pp. 45–52; [Charleston] *South-Carolina Gazette*, September 12, 1771; Walter B. Edgar *et al.* eds., *Biographical Directory of the South Carolina House of Representatives*, 4 vols. thus far (Columbia: University of South Carolina Press, 1974–), 2 (1977), pp. 321–22.

20. Heyward, *Seed from Madagascar*, pp. 62–83; Heyward, "The Heyward Family of South Carolina," pp. 156–57; Edgar *et al.*, eds., *Biographical Directory of the South Carolina House of Representatives*, 3 (1981), pp. 333–35; Manuscript returns, Seventh Census of the United States, 1850, Agriculture, South Carolina, Colleton District, St. Bartholomew's Parish, return for Nathaniel Heyward, Sr., South Carolina Department of Archives and History, Columbia, S.C.

21. Heyward, *Seed from Madagascar*, pp. 62–72, 90; Heyward, "The Heyward Family of South Carolina," pp. 156–57.

22. Heyward, *Seed from Madagascar*, pp. 90–98; Heyward, "The Heyward Family of South Carolina," pp. 208–9.

23. Heyward, *Seed from Madagascar*, pp. 77–78, 90–106; Manuscript returns, Eighth Census of the United States, 1860, Slave Schedules, South Carolina, Colleton District, St. Bartholomew's Parish, return for Charles Heyward, South Carolina Department of Archives and History. One can learn a great deal about Charles Heyward's planting operations from his surviving day books and account books. See Account Books & Day Books, Heyward Family Papers, South Caroliniana Library, University of South Carolina, Columbia, S.C.

24. Heyward, *Seed from Madagascar*, pp. 128–47; Heyward, "The Heyward Family of South Carolina," pp. 208–9.

25. Heyward, *Seed from Madagascar*, pp. 148–49; Heyward, "The Heyward Family of South Carolina," pp. 215–17.

26. Heyward, *Seed from Madagascar*, pp. 148–49; Heyward, "The Heyward Family of South Carolina," pp. 215–17.

27. Heyward, *Seed from Madagascar*, pp. 148–59; David D. Wallace, *The History of South Carolina*, 4 vols. (New York: American Historical Society,

1934) 4 (Biographical Volume), p. 588; Heyward, "The Heyward Family of South Carolina," pp. 215–17.

28. Heyward, *Seed from Madagascar*, p. 153; Duncan Clinch Heyward, Auto-biographical Sketch, December 1936[?], University of North Carolina Press Records, Sub-group 4, D. C. Heyward Correspondence, Manuscripts Department, University of North Carolina, Chapel Hill, N.C.

29. Note that the figure given on page 152 of *Seed from Madagascar* for the amount of rice land on Lewisburg and Amsterdam plantations—8,000 acres—resulted from a typographical error. See Duncan C. Heyward to William T. Couch, January 21, 1938, University of North Carolina Press Records, Sub-group 4, D. C. Heyward Correspondence, Manuscripts Department, University of North Carolina.

30. Heyward, *Seed from Madagascar*, pp. 152–53. At one point, E. B. Heyward wrote that he planned to shift production on his Combahee properties from rice to cotton because of his shortage of capital. With the help of Ohioan John Kilgore and others, however, he was able to gain access to sufficient capital to remain in rice. See E. B. Heyward to Allen C. Izard, July 16, 1866, Heyward Family Papers, Box I, South Caroliniana Library, University of South Carolina.

31. Heyward, Auto-biographical Sketch, December 1936[?], University of North Carolina Press Records, Sub-group 4, D. C. Heyward Correspondence, Manuscripts Department, University of North Carolina; James C. Hemphill, ed., *Men of Mark in South Carolina*, 4 vols. (Washington, D.C.: Men of Mark Publishing Company, 1907–1909), 1: 6–11.

32. See the sources cited in note 31. See also "Duncan Clinch Heyward," in *The National Cyclopaedia of American Biography*, 75 vols. (New York: J. T. White & Co., 1888–1984), 31: 274–75.

33. D. C. Heyward, Sunset at Lamont, January 24, 1881, Heyward Family Papers, Box I, South Caroliniana Library, University of South Carolina. Note that an earlier version of this poem, dated January 21, 1881, exists as well, and is included in Box I of the Heyward Family Papers.

34. D. C. Heyward, Going to War, April 4, 1883, Heyward Family Papers, Box I, South Caroliniana Library, University of South Carolina. For another example of Heywardian verse, see Lines to 'My Sweetheart,' dated April 7, 1883, in Box I of the Heyward Family Papers.

35. Heyward, Auto-biographical Sketch, December 1936[?], University of North Carolina Press Records, Sub-group 4, D. C. Heyward Correspondence, Manuscripts Department, University of North Carolina; "Duncan

Clinch Heyward," in *The National Cyclopaedia of American Biography*, 31: 274.

36. Heyward, *Seed from Madagascar*, pp. 158–59. Note that Izard's middle name is misspelled in *Seed from Madagascar*. See Duncan C. Heyward to William T. Couch, January 21, 1938, University of North Carolina Press Records, Sub-group 4, D. C. Heyward Correspondence, Manuscripts Department, University of North Carolina.

37. Heyward, *Seed from Madagascar*, pp. 44, 45, 201–10; Duncan C. Heyward to Uncle Houstoun [Clinch], April 15, 1887, Heyward Family Papers, Box I, South Caroliniana Library, University of South Carolina.

38. Duncan C. Heyward to Uncle Houstoun [Clinch], April 15, 1887, Heyward Family Papers, Box I, South Caroliniana Library, University of South Carolina; Heyward *Seed from Madagascar*, p. 43.

39. See Account Book, 1887–1915, Duncan Clinch Heyward Papers, South Caroliniana Library, University of South Carolina. Note, however, that a favorable balance between rice receipts and direct production expenses does not necessarily mean that Heyward's operations as a whole were profitable. Many other variables are involved, that is to say, in determining overall rates of return. Note, too, that Heyward planted a small amount of rice in 1913. See Heyward, *Seed from Madagascar*, pp. 246–47.

40. On the poor rates of return on rice, even in the antebellum period, see Coclanis, *The Shadow of a Dream*, pp. 140–41 especially; Dale E. Swan, *The Structure and Profitability of the Antebellum Rice Industry 1859* (New York: Arno Press, 1975), Preface, and pp. 75–84. Heyward, too, views the cultivation of rice in the low country as increasingly unsound in economic terms in the antebellum period. See Heyward, *Seed from Madagascar*, pp. 83–84.

41. Giuseppe Tomasi di Lampedusa, *The Leopard*, trans. Archibald Colquhoun (1958; rpt. New York: Pantheon, 1960). On the inertial tendencies of hydraulic crops such as paddy rice, see Coclanis, *The Shadow of a Dream*, pp. 140–43. Note that despite its declining profit possibilities, rice still may have been the low country's best economic option in the late nineteenth century. On this point, see Coclanis, *The Shadow of a Dream*, pp. 140–58. Heyward seems to have recognized the point as well. See Heyward, *Seed from Madagascar*, p. 247.

42. Heyward, "The Heyward Family of South Carolina," p. 223; Heyward, Auto-biographical Sketch, December 1936[?], University of North Carolina Press Records, Sub-group 4, D. C. Heyward Correspondence, Manuscripts Department, University of North Carolina; Hemphill, ed., *Men of Mark in South Carolina*, 1: 9.

43. Warranty Deed [for 3 lots in Highland Park, "an addition to Sioux City, Iowa"], Leighton Wynn, Trustee of Woodbury County, Iowa, to Duncan C. Heyward, October 27, 1887, Heyward Family Papers, Oversize Material, South Caroliniana Library, University of South Carolina; Temple Harris to Duncan C. Heyward, November 27, 1889; Temple Harris to Duncan C. Heyward, March 7, 1891, Heyward Family Papers, Box I, South Caroliniana Library, University of South Carolina; Heyward, *Seed from Madagascar*, pp. 201–10.

44. Hemphill, ed., *Men of Mark in South Carolina*, 1: 9; "Duncan Clinch Heyward," in *The National Cyclopaedia of American Biography*, 31: 274; Margaret Ola Spigner, "The Public Life of D. C. Heyward, 1903–1907" (M.A. thesis, University of South Carolina, 1949), pp. 1–17; Ernest M. Lander, Jr., *A History of South Carolina, 1865–1960* (Chapel Hill: University of North Carolina Press, 1960), pp. 30–55, especially p. 46; [Columbia] *The State*, August 23, 1903; *The State*, January 17, 1905.

45. See Spigner, "The Public Life of D. C. Heyward, 1903–1907," passim; Lander, *A History of South Carolina, 1865–1960*, pp. 47–49. On progressivism in South Carolina, see Robert M. Burts, *Richard Irvine Manning and the Progressive Movement in South Carolina* (Columbia: University of South Carolina Press, 1974); David L. Carlton, *Mill and Town in South Carolina, 1880–1920* (Baton Rouge: Louisiana State University Press, 1982); Sandra Corely Mitchell, "Conservative Reform: South Carolina's Progressive Movement, 1915–1929" (M.A. thesis, University of South Carolina, 1979). On the Bourbons in South Carolina, see William J. Cooper, Jr., *The Conservative Regime: South Carolina, 1877–1890* (1968; rpt. Baton Rouge: Louisiana State University Press, 1991).

46. Frank E. Jordan, Jr., *The Primary State: A History of the Democratic Party in South Carolina 1876–1962* (Columbia: Democratic Party of South Carolina, n.d.), pp. 22–24; Spigner, "The Public Life of D. C. Heyward, 1903–1907," pp. 9–17, 67–71.

47. Spigner, "The Public Life of D. C. Heyward, 1903–1907," pp. 18–66; Lander, *A History of South Carolina, 1865–1960*, pp. 47–49, 53–55, 66–69. On the Greenwood lynching, see Harry Legare Watson to Ella D. Watson, August 17, 1906, John Dargan Watson Papers, South Caroliniana Library, University of South Carolina; *The State*, August 17, 1906; Duncan C. Heyward, Ms. [Account of Greenwood Lynching], n.d. [c.1932], Heyward Family Papers, Box II, South Caroliniana Library, University of South Carolina.

48. Heyward, Auto-biographical Sketch, December 1936[?], University

of North Carolina Press Records, Sub-group 4, D. C. Heyward Correspondence, Manuscripts Department, University of North Carolina; Hemphill, ed., *Men of Mark in South Carolina*, 1: 11; Wallace, *The History of South Carolina*, 4: 588; *Who Was Who in America*, 10 vols. thus far (Chicago: The A. N. Marquis Co., 1943–   ), 2 (1943–1950): 251.

49. Heyward, *Seed from Madagascar*, pp. 239–50; Heyward, Auto-biographical Sketch, December 1936[?], University of North Carolina Press Records, Sub-group 4, D. C. Heyward Correspondence, Manuscripts Department, University of North Carolina; Wallace, *The History of South Carolina*, 4: 588; *Who Was Who in America*, 2: 251; Robert Sobel and John Raimo, eds., *Biographical Dictionary of the Governors of the United States, 1789–1978*, 4 vols. (Westport, Conn.: Greenwood Press, 1978), 4: 1426–27.

50. Duncan C. Heyward to William T. Couth, August 22, 1936, University of North Carolina Press Records, Sub-group 4, D. C. Heyward Correspondence, Manuscripts Department, University of North Carolina.

51. Elizabeth Allston Pringle (Patience Pennington), *A Woman Rice Planter* (New York: Macmillan Co., 1913). In 1992 a new edition of the book was published in the Southern Classics Series of the University of South Carolina Press. The new edition contains an excellent introduction by Charles Joyner.

52. On Heyward's feelings about abandoning rice production, see *Seed from Madagascar*, pp. liii–liv, 211, 249–50.

53. Gordon G. Sikes to Duncan C. Heyward, February 25, 1920; H. J. A. Lawton to Duncan Clinch Heyward, December 20, 1926; Brigadier General James McKinley to Duncan C. Heyward, April 3, 1931, Heyward Family Papers, Box II, South Caroliniana Library, University of South Carolina; *The State*, February 27, 1927; March 27, 1927; April 3, 1927; April 24, 1927; May 1, 1927; May 29, 1927; July 3, 1927; July 10, 1927; July 31, 1927; August 21, 1927; August 28, 1927; September 4, 1927. These sketches appeared in two other papers as well. See Heyward, Auto-biographical Sketch, December 1936[?], University of North Carolina Press Records, Sub-group 4, D. C. Heyward Correspondence, Manuscripts Department, University of North Carolina.

54. Clements Riley to Duncan C. Heyward, January 30, 1935; William T. Couch to J. T. Gittman, September 9, 1936; Duncan C. Heyward to James C. Verieux, October 6, 1936, Heyward Family Papers, Box II, South Caroliniana Library, University of South Carolina. Note that Ripley wrote to both Bobbs-Merrill and Harcourt, Brace on Heyward's behalf, but, apparently, neither publisher was interested in Heyward's manuscript. On William Terry

Couch and the early history of the University of North Carolina Press, see Daniel J. Singal, *The War Within: From Victorian to Modernist Thought in the South, 1919–1945* (Chapel Hill: University of North Carolina Press, 1982), pp. 265–301.

55. William T. Couch to Duncan C. Heyward, August 20, 1936; Robert B. House to William T. Couch, September 7, 1936; J. G. deRoulhac Hamilton to William T. Couch, September 22, 1936, University of North Carolina Press Records, Sub-group 4, D. C. Heyward Correspondence, Manuscripts Department, University of North Carolina. Also see J. G. deRoulhac Hamilton to Duncan C. Heyward, September 22, 1936, Heyward Family Papers, Box II, South Caroliniana Library, University of South Carolina.

56. See, for example, William T. Couch to Duncan C. Heyward, October 15, 1936; William T. Couch to Carl T. Julien, November 4, 1936; Carl T. Julien to Duncan C. Heyward, November 10, 1936; William T. Couch to Duncan C. Heyward, December 10, 1936; William T. Couch to Duncan C. Heyward, December 14, 1936, Heyward Family Papers, Box II, South Caroliniana Library, University of South Carolina. See also Duncan C. Heyward to William T. Couch, December 12, 1936, University of North Carolina Press Records, Sub-group 4, D. C. Heyward Correspondence, Manuscripts Department, University of North Carolina.

57. *New York Times*, October 19, 1937; *American Historical Review* 44 (January 1939): 463. Heyward received many letters from friends and others telling him that they had enjoyed *Seed from Madagascar*. For a sampling of such letters, see Heyward Family Papers, Box II, folders 5, 7, 8, 9, 10, 11, South Caroliniana Library, University of South Carolina. Many of the letter writers, it should be noted, had received complimentary copies of the book.

58. William T. Couch to Carl T. Julien, November 4, 1936, Heyward Family Papers, Box II, South Caroliniana, University of South Carolina; William T. Couch to Ander Braun, July 20, 1937; J. G. deRoulhac Hamilton to William T. Couch, September 22, 1936, University of North Carolina Press Records, Sub-group 4, D. C. Heyward Correspondence, Manuscripts Department, University of North Carolina. Note that Gittman's Book Shop was located at 1227 Main Street in downtown Columbia.

59. Heyward, *Seed from Madagascar*, Total Sales to 6/30/48, Ms., University of North Carolina Press Records, Sub-group 4, D. C. Heyward Correspondence, Manuscripts Department, University of North Carolina.

60. Porter Cowles to Katherine B. Heyward, February 1, 1961, University of North Carolina Press Records, Sub-group 4, D. C. Heyward Correspondence, Manuscripts Department, University of North Carolina.

61. In the first edition of *Seed from Madagascar* there were sixty-four photographs by Julien interspersed in the text. On Julien, see John Hammond Moore, "Carl Julien: The Eye of South Carolina," *Carologue* 7 (Winter 1991): 6–7, 14–17. Julien's photographic illustrations appeared in many books with South Carolina themes, but his most successful single work was probably his collaborative effort with Chapman J. Milling, *Beneath So Kind a Sky*. This book of photographs of South Carolina's natural and built environments, with a brief text by Milling, was published by the University of South Carolina Press in 1947 and is now in its eighth printing.

62. On these last two questions—slave property-holding and ethnicity—see Heyward, *Seed from Madagascar*, pp. 54, 180–85, 197. Note that Heyward could understand and speak Gullah and the passages in Gullah in *Seed from Madagascar* are rendered relatively accurately. On the basic rules of the Gullah language, see Charles Joyner, *Down by the Riverside: A South Carolina Slave Community* (Urbana: University of Illinois Press, 1984), pp. 196–224.

63. See, for example, Heyward, *Seed from Madagascar*, pp. 57, 60, 89, 106, 118.

64. Heyward, *Seed from Madagascar*, pp. 154, 157, 162.

65. Heyward, *Seed from Madagascar*, pp. 25–26, 182.

66. Heyward, *Seed from Madagascar*, pp. 249–50.

67. J. G. deRoulhac Hamilton to William T. Couch, September 22, 1936, University of North Carolina Press Records, Sub-group 4, D. C. Heyward Correspondence, Manuscripts Department, University of North Carolina; August Kohn, *The Cotton Mills of South Carolina* (Columbia: State of South Carolina Department of Agriculture, Commerce and Immigration, 1907), p. 53. To Mills Lane, Chairman of the Board of the Citizens and Southern National Bank in Savannah, Heyward was also at one time "the 'handsomest' man in South Carolina." See Mills B. Lane to S. F. Clabaugh, July 31, 1937, Heyward Family Papers, Box II, South Caroliniana Library, University of South Carolina.

68. For Heyward's comment that he considered the story he was telling a "tragic" one, see *Seed from Madagascar*, p. 3.

TO
THE MEMORY OF
MY WIFE

MARY CAMPBELL HEYWARD

WHOSE COURAGE NEVER FAILED
DURING MY LONG STRUGGLE
AS A RICE PLANTER

I am grateful to the following friends who encouraged me to write this book: Mr. James T. Gittman, *of Columbia, South Carolina,* Mr. James C. Derieux, *of Summerville, South Carolina, and* Dr. George A. Wauchope, *of the University of South Carolina. I am also grateful to my friend,* Mr. A. S. Salley, *State Historian, for helping me verify certain facts, and to my daughter-in-law,* Beulah H. Heyward, *for her time and labors in typing the original manuscript.*

D. C. H.

COLUMBIA, S. C.
April 23, 1937

FOREWORD

OVER the mantel in my dining room there hangs the portrait of my great-grandfather, Nathaniel Heyward, the largest rice planter of his day. When I look at this portrait, I imagine that the eyes regard me with sadness and disapproval, because I am no longer a rice planter.

Nathaniel Heyward owned and planted seventeen plantations on the Combahee River in South Carolina, some of which his father and he had reclaimed. Three of these he always valued above all the others, speaking of them as his "gold mines," and many years later I owned and planted all three. They were among the very last to be planted in rice in South Carolina, and I was the last to plant them. When the end of the industry came on the southern Atlantic seaboard, they passed into other hands, and never since has a crop of rice been grown there. I have the feeling that "Ole Maussuh," as his Negroes called him, must know of this loss and that he finds it hard to understand, for never in his long experience as a planter did he make a deed to a plantation or sell one of his slaves.

Why, then, are his "gold mines" no longer owned by one of his name? This is the question he seems to be asking me, and I shall try to answer it in the pages that follow. Also, should it come to pass in the hereafter that I meet him face to face and he reproach me for losing the lands he valued so highly, I shall welcome the chance to prove to him that I was not to blame, even though his eyes then look more sternly at me than they do now in the old portrait.

I shall give as accurately as I can an account of the rice industry as it existed in the coastal section of South Carolina and Georgia, although I shall write only of the Combahee River, typical of all the tidewater swamp plantations, which were among the first to be reclaimed and the last to be planted on the Atlantic seaboard. On all these plantations conditions were practically the same as to methods of planting and management of Negro labor.

Until the Civil War there had lived in South Carolina scarcely a man of my name who was not a rice planter, or an owner of a rice plantation. The reclamation of the alluvial swamps and their cultivation necessitated the owning of slaves, and from colonial days my ancestors were among the largest slaveholding families in the South. Certain plantation records in regard to the slaves owned by my immediate family have long been in my possession, and my special interest in them has arisen from the fact that I knew a number of these Negroes long after they were freed. When I knew them, they were living in the same settlements where they had been born and reared; they were planting the same crop they had always planted, and by practically the same methods; they were under an overseer who for ten years had managed them as slaves. As far as these Negroes were concerned, a pay day was almost the only difference between my grandfather's time and mine.

FOREWORD

In the pages that follow, I shall confine myself to stating what I know of the system of slavery. I have had many opportunities to learn the true conditions, having associated with former slaveholders and having known intimately men who, as overseers, had come in close contact with slaves. And I have also asked many questions of old Negroes relating to their former servitude.

SEED FROM MADAGASCAR

1

CAROLINA GOLD RICE

OFTEN during my years as a planter, when the rice industry on our South Atlantic coast was rapidly being abandoned, I have sat under a great cypress, growing on my river bank, and, looking across the broad expanse of my rice fields, have thought of their strange and remarkable history. There would come to my mind the great and fundamental changes, racial and social as well as economic, which have taken place in a short space of time. And I have wished that the old tree above me could tell the tragic story of those fields, recounting events of which it had been a silent witness.

If the tree could only have spoken, I know its story would have begun at a time when the swamp, on whose edge it grew, was the favorite hunting ground of the red man. It would then have told of the coming of the white man, who drove the red man far away and took from him his lands. Next it would have told how the black man came, brought from far across the sea, how he felled the trees in the swamps and cleared them of their dense undergrowth, letting the sunshine in; then, how he drained the lowlands and grew crops of golden grain; and finally it would have told of the emancipation of the black man, who, after years of servitude, worked on faithfully as a freedman.

The rest of the story of my rice fields would for me have needed no telling. It would have dealt with the years when the white man was compelled, by conditions beyond his control, to give up planting, and the black man moved away seeking employment elsewhere, leaving fertile lands, the only naturally irrigated ones in this country, to revert to their former state.

The story of our former rice fields begins more than two centuries ago.

Carolina Gold rice,[1] world renowned because of its superior quality as compared with all other varieties of rice throughout the world, was grown from seed brought to the province of Carolina about the year 1685. This rice had been raised in Madagascar, and a brigantine sailing from that distant island happened, in distress, to put into the port of Charles Town. While his vessel was being repaired, its captain, John Thurber, made the acquaintance of some of the leading citizens of that town. Among them was Dr. Henry Woodward, probably its best known citizen, for he had the distinction of being the first English settler in the province. He had accompanied Sandford on his first exploring expedition and had volunteered to remain alone at Port Royal in order that he might study the language of the Indians and familiarize himself with the country, in the interests of the Lords Proprietors.

To Woodward Captain Thurber gave a small quantity of rice—less, we are told, than a bushel—which happened to be on his ship. "The gentleman of the name of Woodward," to quote the earliest account of this occurrence, "himself

[1] So called on account of the color of its outer hull.

planted some of it, and gave some to a few of his friends to plant." [1]

Thus it came about that I, on the distaff side a descendant of Dr. Woodward, seem to have been destined to spend the best years of my life in seeking to revive an industry in the pursuit of which four generations of my family had been successful.

Three states bordering on the South Atlantic, North Carolina, South Carolina, and Georgia, are the only states in this country where for upwards of two centuries rice was grown. Nearly all of it, however, was produced in South Carolina and Georgia, partly because of a slightly warmer climate, but mainly because of the numerous tidal rivers which flow through these states and empty into the ocean. Along the Cape Fear River in North Carolina, it must be admitted, the finest quality of rice was grown, and for many years seed raised there was sold to the planters farther south, in order to preserve the quality of their rice.

The principal rivers in South Carolina, along which rice was planted, were the Waccamaw, the PeeDee, the Santee, the Cooper, the Edisto, and the Combahee. There were also many large rice plantations on the Savannah River, which separates South Carolina and Georgia. Farther south in Georgia were the Ogeechee, the Altamaha, and the Satilla rivers, the last near the Florida line. Some of these rivers are long, having their sources in the mountains, while others are much shorter. All of them are affected for a number of miles by the rise and fall of the tide, the result being that the fresh

---

[1] A. S. Salley, "The Introduction of Rice Culture into South Carolina," *Bulletin of the Historical Commission of South Carolina*, No. 6.

water they contain is backed up in the rivers and then drawn down again as the water in the ocean rises and falls. Great salt-water marshes lie on either side of the rivers as they approach the ocean, while higher up they were originally bordered by dense cypress, gum, and cedar swamps where the water was fresh, though rising and falling with the tide. It was in these fresh-water swamps that rice was successfully grown for the longest period of years.

As early as August 31, 1663, Lord Albemarle, one of the Lords Proprietors, wrote to the Governor of the Barbadoes, advising the planters there to settle in the proposed province of Carolina. Among the inducements offered he suggested the growing of rice. In his letter he said, "The commodyties I meane are wine, oyle, reasons, currents, rice, silk, etc." Of these, rice alone was destined to be successfully produced over a long period of years.

The plans of the Lords Proprietors regarding the growing of rice in the new province took practical shape in 1672, for in that year they had a barrel of rice sent to Charles Town in a vessel named the "William and Ralph"; and one must assume that this rice was intended for seed. By 1690 some headway had been made in the growing of rice, for the leading men of the province petitioned Governor Sothell to arrange with the Lords Proprietors that the people be allowed to "pay their quit rents in the most valuable and merchantable produce of their land," among which they included rice. These products, they reported, were "naturally produced here."

Also during this period the General Assembly of the province had ratified acts to protect those who should perfect labor-saving machinery for the purpose of husking and cleaning rice. A few years later the Assembly protested against an export duty on rice. How different was this attitude of the early rice planters from that of the planters of my day! We wanted an import duty. Though Southerners, we favored a tariff on rice, for we sorely needed protection. However, we never went so far as to claim that ours was an "infant industry."

By the year 1700 there was being produced in the province more rice "than we had ships to transport," according to the Governor and Council. Edward Randolph, collector of customs for the Southern department of North America that year, wrote: "They have now found out the true way of raising and husking rice. There has been above three hundred tons shipped this year to England besides about thirty tons to the Islands." This progress must have caused the Lords Proprietors to be exceedingly optimistic as to the future of rice cultivation in Carolina, for they congratulated themselves upon "what a staple the province of Carolina may be capable of furnishing Europe withall," and added that "the grocers do assure us it's better than any foreign rice by at least 8s the hundred weight."

At any rate, for the praise so soon bestowed by the grocers of London on the quality of the rice exported from the province of Carolina, and for the demand from abroad for this variety of rice, and also for the success which for upwards of two centuries attended the growing of rice in South Carolina and Georgia, we are principally indebted to Captain Thurber. Had it not been for him, the once celebrated Carolina Gold rice probably would never have been planted in America.

Another variety of rice later planted in South Carolina was known as Carolina White rice. This rice made a beautiful sample when prepared for market, and could scarcely be distinguished from the Gold rice, but its tendency to shatter when being harvested, if slightly over-ripe, caused it to be planted only to a limited extent.

It has never been known definitely from what country this Carolina White rice was first imported, but many of our rice planters believed it had been brought from China, where for unnumbered centuries rice has been grown and where there are today numerous varieties. Were I to hazard a guess as to who was responsible for bringing this white rice to the province, I should name Robert Rowand, born in Glasgow in 1738, who, when a boy, came to Charles Town. He later purchased an inland swamp plantation, on Rantowles Creek, about twenty-five miles southwest of Charles Town. He must have succeeded as a planter, for seemingly he was a man of means and traveled extensively. There is every reason to believe that he visited China, and that, while there, he had certain attractive pictures painted, illustrating the way the Chinese planted rice in those days, and showing the implements used in the process.

A few years ago in the town of Summerville, South Carolina, at the home of a friend, the late Mr. S. Lewis Simons, I saw these pictures. (They have since been destroyed by fire.) There were quite a number of them, and I am sure there are no others like them anywhere in this country. In size, they were about eighteen by fourteen inches, exceedingly well executed in bright water colors. They showed in great detail the growing of rice in China, from the preparation of the ground to the gathering and pounding of the grain. The pictures were evidently painted by a Chinese artist. The figures of men and women, the landscape, with its green trees here and there, blue mountains and hills rising out of the level plains intersected by canals, were entirely Oriental. On the front of the cover was written, "Painted in China prior to the Revolutionary War for one of the first South Carolina Rice Planters, Illustrating the Chinese Method of Cultivating Rice." The rice planter referred to was undoubtedly Robert Rowand, and the pictures had come into the possession of Mr. Simons through his wife, a Miss Mayrant, the Mayrant family having been rice planters for several generations. Mrs. Simons was a lineal descendant of Robert Rowand on her maternal side through the Drayton family.

The first of these Chinese pictures shows the plowing and harrowing of the soil of the seed-beds, both processes being done under water, the Chinaman and his black "water buffalo" nearly up to their knees, and the latter looking as if he did not trust his footing and was anxious to turn back. Then follows the sowing of the seed broadcast on the water, the transplanting of the rice by hand in the fields, the cultivating of the growing crop, the harvesting, until finally the rice is carried to the barnyard, where it is threshed and pounded. Anyone at all familiar with the methods used by the early planters in South Carolina cannot fail to be struck by their similarity to the methods shown in the old Chinese pictures, and especially is this true of the implements used. With the exception of sowing the rice in the water and transplanting it, nearly every picture recalled our system of planting. Many of the implements were almost identical with ours. There were the flail-sticks, being used in exactly the same way; the sickle, only a little straighter than ours, with which the rice was cut; the mortar and pestle and hand-fans; the "boards," as the Negroes used to call them, with which, indoors on a floor, the rice was pushed from place to place; and also the baskets in which the rice was carried.

These paintings of Robert Rowand's convinced me that our early methods of rice culture were adopted largely from the Chinese. For if, instead of the Chinese settings of the pictures, if in the place of the men with queues and black slanting eyes, dressed in bright-colored costumes, there could have been substituted the scenery of our rice fields and the Low Country Negroes at work, I could readily have believed the scenes were laid in our Black Border instead of in that far Eastern land.

2

INLAND SWAMP PLANTATIONS

THE rice industry in South Carolina can be divided into two
fairly well defined periods, which differed not only in the
location of the plantations but also in the method of irrigation used by the planters. During the first period, beginning
in the latter part of the seventeenth century and continuing
until the middle of the eighteenth, rice was grown on inland
swamps. During the second period, beginning in the middle
of the eighteenth century and continuing until the end of the
industry, when an ample supply of slave labor could be obtained, the planting of rice on inland swamps was gradually
abandoned and its cultivation transferred to the extensive and
thickly timbered swamps which bordered the fresh-water
tidal rivers.

The inland swamps, the lands first reclaimed, were near
the white settlements along the coast, for in those early days
the interior of the province was still occupied by the Indians,
who were not friendly, and the safety of the white man had
to be considered. The inland swamps reclaimed were small
in area and drained into salt-water rivers and creeks. As the
rivers of South Carolina and Georgia approach the ocean,
there are numerous creeks which empty into them and into
some of the bays along the coast. These creeks have their

sources in inland swamps, which are often quite extensive, with highlands on either side. As the creeks near the rivers, they wind through salt-water marshes, and many of them are wide and deep, with a rise and fall of the tides from four to six feet. Higher up in the swamps, however, the water is nearly always fresh, for the rainfall from an extensive territory drains into them and the level of the land slopes gradually toward the sea.

Salt water, even brackish water, is fatal to growing rice, while fresh water and a certain amount of drainage are essential to its successful cultivation. Hence the early rice planters in the province of Carolina, in order to grow rice in inland swamps, had two problems to solve. Their fields had to be adjacent to salt water in order to get the necessary drainage; and, to irrigate their rice when needed, they had to have an ample supply of fresh water. Difficult as such problems might appear, the planters soon adopted a plan which, though by no means perfect, they pursued for many years, with fair success.

To reclaim an inland swamp the first work to be done was to throw up a strong earth dam across its lower end. The purpose of this dam was to prevent salt water from overflowing the parts of the swamp to be planted. Then, higher up in the swamp, smaller dams were built. The land between these dams was known as "squares," and each square was given a name by which it could be designated. All of the dams extended entirely across the swamp from the highland on one side to the highland on the other.

Through the dam at the lower end of the swamp one or more large sluice gates were placed. These sluice gates were known as "trunks," a name brought to the province by the early English settlers, who had seen them used in the fresh-water marshes of England. These trunks were long wooden boxes made of thick plank, with a door at each end. The doors were hung on uprights and were capable of being raised and lowered automatically with the rise and fall of the tide. When the door on the river end of a trunk was lowered and the door on the field end raised, the water in the field at low tide passed through the trunk, swinging the door open, but when the tide rose, its pressure against the outer door caused it to shut tightly, and no water was admitted into the field. When it was desired to flood the field with salt water, which was scarcely ever the case on inland swamp plantations, unless for the purpose of shooting ducks in the fall and winter months when the crop had been harvested, all that was necessary was to reverse the raising and lowering of the doors.

When the dams had been built and the trunks installed, the clearing of the swamp was begun. This was not, in most instances, a great undertaking, for very large trees seldom grew in the lower portions of these swamps, nor was the undergrowth very dense. When the land was cleared, canals and ditches were dug. This also was not difficult work, for the dark, alluvial soil yielded readily to the shovel. By means of these ditches the lands to be planted were drained to the greatest possible extent. The smaller of the ditches ran across the swamp, and were known as "quarter" ditches, while the larger, running in both directions, were called "face" ditches. These names continued to be used during the life of the industry in South Carolina and Georgia.

Nearly equal in size to the large dam at the lower end of the swamp was another dam, the highest up in the swamp. This dam held the water in the upper unreclaimed portion of the swamp and made it a reservoir, to be used for irrigation. These reservoirs were, however, most uncertain, for the amount of water they contained was dependent upon rainfall, and a long dry season meant the failure of a crop.

Nathaniel Heyward, my great-grandfather, never planted an inland swamp plantation but one year in all of his life, and that one year fully satisfied him. This venture of his occurred just one hundred and fifty-eight years ago, and was his first rice-planting experience. He had made a fine crop, but in August a freshet came down the swamp when his rice was "in blossom" and completely destroyed it. He drove to the place the next morning, took one look, not at the crop, for he couldn't see that, but at the water covering it, got into his buggy, and drove away, never to set foot on the place again. Instead, he moved over on the Combahee River on lands which had belonged to his father, but were then owned by his two brothers. He reclaimed many acres along that tidal river, bought many plantations, and became the largest rice planter of his day.[1]

It was principally this lack of water at one time and too much water at another that caused, in later years, the inland swamp plantations to be gradually abandoned, and the cultivation of rice transferred to the much larger swamps adjacent to fresh-water rivers, in which the rise and fall of the tides could be depended upon for irrigation and drainage.

[1] Life of Nathaniel Heyward, by his grandson, Dr. Gabriel Manigault, of Charleston, South Carolina, written 1895, but not published.

In the Low Country of South Carolina there is told a story that in the early years of rice planting, during a severe drought, the minister in charge of Sheldon Church (built 1745–1755) in what is now Beaufort County, was asked to offer a prayer for rain. Complying with the request, and following his prayer book, he prayed the Lord to send "such moderate rains and showers that we may receive the fruits of the earth," whereupon an old rice planter rose in his pew and walked out of the church, muttering as he did so, "What in hell do I want with moderate rains and showers? It will take regular trash-moving, gully-washing rain to do my crop any good."

Many of the leading citizens and statesmen in pre-Revolutionary days, including at least three of the four South Carolina signers of the Declaration of Independence, were rice planters, and planted inland-swamp plantations. They had fine residences on their plantations, but not one of the old houses now remains. A stranger today traveling along a former stage road in the lower part of South Carolina would never suspect that he was passing through one of these old places, for the swamps where rice was first planted have now grown up again in trees. When he drives through the swamp and gets on the highland, his attention is often attracted by an avenue of live oaks leading through the woods, the trees hoary with age. At the head of the avenue and close to the main road he sees two brick gateposts on which no gate has swung for years. Looking up the avenue he will see long streamers of gray moss hanging from the live oaks, making them look even older than they are. Their far-reaching limbs and their decaying condition plainly tell how long it has been since the avenue was laid out and the trees planted in such even rows.

These avenues always lead to the place where the dwelling of the rice planter once stood, but there is now no sign of it; every building has either been burned or has rotted down. Only a few brickbats mark where the chimneys of the house stood. On some of these places an old brick wall, enclosing a family burying ground, is still in good condition.

In the yard which once surrounded the dwelling an old magnolia tree, which has withstood many a storm, may yet be seen, and farther away toward the foot of a high bluff a few scrub palmettoes, looking lonely and desolate off by themselves. Nevertheless, despite the solitude of the place and its profound stillness, a visitor, on a clear day, will be enchanted by the view from the bluff. Beneath a sky wonderfully blue, a broad expanse of salt-water marshes spreads out before him, and through it flows a river or creek, its slow-moving water sparkling in the sunshine. As he looks across the marshes, their varying shades of green are usually stirred by a breeze coming inland through an opening on the horizon, toward the southeast, where the ocean lies.

3

RECLAIMING FRESH-WATER
TIDAL SWAMPS

A CENTURY and a half ago, on his memorable visit to South Carolina and Georgia, though he had never before seen a rice plantation, Washington was much impressed with the adaptability of our large fresh-water tidal swamps along the rivers for the cultivation of rice.

On his journey through South Carolina the President spent a night at the home of Captain William Alston on the Waccamaw River near Georgetown, and on April 29, 1791, wrote in his diary:

Captain Alston is a gentleman of large fortune and esteemed one of the neatest planters in the State of South Carolina, and a proprietor of the most valuable ground for culture of rice. His house which is large, new and elegantly furnished, stands on a sand hill, high for the country, with its rice fields below, the contrast of which, with the lands back of it, and the sand and piney barrens through which we passed, is scarcely to be conceived.

And the Virginia farmer also quickly recognized the advantages for growing rice, which the fresh-water tidal swamps had over the inland swamps, for he added this comment:

The rice planters have two modes of watering their fields, the first by the tide, the other by reservoirs, drawn from adjacent lands. A crop without either is precarious because a drought may not only impair but destroy it.

At the time of the President's visit a large portion of Captain Alston's fields probably were flooded, and this must have been novel to Washington, for nowhere else in this country was such a method of irrigation used. But I doubt if the sight of so much water appealed to a highland farmer. I remember on one occasion when my fields were deep under water as the result of a bad break in my river bank, and I was being rowed over them after dark, I asked a young farmer from Tennessee, who was in the boat, if he had ever seen so much water, and his reply was: "No, not on the farm." As there was from one to two feet of water on my farm, and the break had to be stopped before it could be drained off, I did not appreciate his reply.

The reclamation of the fresh-water swamps was a great undertaking, considerably greater than that of the inland swamps which bordered on the salt-water marshes. There were many large white gum, cedar, and cypress trees, and the dark alluvial soil was so soft that one could scarcely walk any distance upon it. To avoid sinking he would have to step from one root to another, or trust his weight to some treacherous tussock. Everywhere his progress was impeded by dense undergrowth, and his clothes and flesh torn by briars. Half the time the land was under water, which slowly receded when the tide in the river ebbed, but which soon returned.

Numbers of alligators were in the swamps in those days. Some of them were very large. On warm days they could often be seen swimming in the river, as they did in my day, with only their eyes and noses showing above the water, or sunning themselves, full length, upon mud flats when the tide was low. Turtles, innumerable during spring and summer, crawled from their holes and lay in single file on fallen logs, ready at the slightest alarm to slide quickly back into the dark water. Snakes of many kinds, especially water moccasins, luxuriated where the sun found its way through the thick foliage, but they were not greatly dreaded by the slaves when they worked in the swamps, nor by the Negroes of later years.

These river swamps varied greatly in width. Some extended a mile or more from the river. Looking across them from the highland in summer, before they were cleared, one saw a vast expanse of green foliage, but when winter came, quantities of moss clinging to the trees became visible, and with the gray bark of the gum trees gave the whole scene a melancholy appearance.

The swamps always extended from the river back to the highland, and at intervals the highland, in rather narrow strips, ran down to the river's edge, forming bluffs, which were not only excellent sites for threshing mills and barnyards, but were also good places for schooners to lie when discharging and loading their cargoes. These strips of highland also often divided adjacent plantations.

The first step in reclaiming the swamp lands was to build a bank along the edge of the river, with both ends joined to strips of highland where they approached the river's edge, and through the bank to place trunks, similar to those used in the inland swamps, for the water to pass through.

When the bank had been built and the trunks installed, the digging of canals and ditches in the swamp followed. Then the trees and undergrowth had to be removed, the greatest undertaking of all. The trees were cut down and burned, but their stumps were never completely removed. Some old cedar stumps, level with the ground, are there to-day and are very hard to get rid of. A modern stump-puller would be useless, for, instead of pulling the stumps up with their long lateral roots, the puller would sink into the soil.

It is hard for us to appreciate in this day of machinery, and especially of dredging machines, what a heavy undertaking it was, by manual labor alone, to reclaim the large river swamps in lower Carolina and transform them into fields adapted to the growing of rice. Even when this work was accomplished, by no means did the early planter's troubles end; for many, they had only begun. All were constantly harassed by the advice and demands of their greedy Lords Proprietors, who, fortunately, were several thousand miles away. All of the plantations were threatened with tropical storms, and some with freshets. The only laborers to be obtained were thoroughly inefficient and took no interest in the work. At first, the credit of the planters was limited, and for some it was difficult to obtain supplies for their slaves. I have thought that they often might have likened themselves to the man of whom the Prophet Amos tells, who "did flee from a lion and a bear met him; and he went into the house and leaned his hand on the wall, and a serpent bit him."

After their crops had been grown and stacked in the barnyards, the planters still had difficult problems to solve. The first was how to remove the grains of rice from the straw; and the second, which required a longer time to solve, was how to find a satisfactory method of preparing the grain for market. To thresh the rice, the planters had, for many years, to resort to flail-sticks. The bundles of rice were placed in rows on the ground, with the heads joining each other. The Negroes walked between the rows, swinging the flail-sticks above their heads, and bringing them down on the heads of rice, thus beating off the grain.

The use of the flail-stick was slow work, especially as it could be carried on only during good weather. When the grain was threshed off, it had to be gathered up and carried to what was known as a "winnowing house," a building about twenty feet off the ground, supported on posts. In the floor of the winnowing house, a grating was placed through which the rice was dropped to the ground, so that, if any breeze were stirring, it would blow away the light and unfilled grains and also any short pieces of straw which were mixed with the rice.

After the rice had passed through these two processes resort was made to the ancient method of pounding rice, that of the wooden mortar and pestle, in order to remove the outer hull, and to do this required an even longer time than did threshing and winnowing. It was thus often far into the winter before the crop could be sent to market. In order to prevent these delays, the planters, very soon after the beginning of the industry, adopted an experiment never before tried in this country—the use of the tides as a source of power. While it could not be pronounced an effective method, it was used for many years.

On practically every abandoned rice plantation in South Carolina can be seen today a square known as the "Mill Pond" square, for it was into these squares that the tides were allowed to flow and ebb during the fall and winter months, when the old pounding-mills were in operation. Water was let into these squares through a brick raceway, the sides of which can often still be seen, and as it flowed into the field and out, it turned an undershot wooden wheel, causing two large stones to revolve, one upon the other, between which the hulls of the rice were rubbed off. The gate to the raceway was not opened until half tide, either flood or ebb, and hence the mill could be operated for only a limited time—for about five hours during the day and the same length of time at night. A simple mechanical rearrangement of the driving belt reversed the direction of the mechanism with the change of the tide.

No real advance, however, was made in the preparation of rice for market until the year 1787, when Jonathan Lucas, an Englishman, then a citizen of Charleston, and a thoroughly educated wheelwright, invented improved machinery for pounding and cleaning. This he first installed in a water-propelled mill on a plantation on the Santee River, after which he built a number of mills and equipped them with his new inventions.

In 1801 Lucas built the first pounding-mill operated by steam, and, as a result of this, use of tides was gradually abandoned, steam engines being used in their stead, and steam threshing mills, which later were built on every rice plantation, replaced the flail-stick.

By some it has been thought that the pounding-mill built by Lucas in 1787 was the first pounding-mill in South Carolina to be operated by the tides, but this is not true. As evidence, one has only to visit my great-great-grandfather's plantation, where it can be seen that he pounded his rice in a mill operated by the tides, for there is the old raceway through which the tides ebbed and flowed, and also the two large stones between which the rice was hulled. Lucas probably was still in England when Daniel Heyward died.

Great credit, however, must be given to Lucas for the improved machinery he invented and the aid he rendered the rice planters in solving one of their most serious problems. His name will always be associated with the milling of rice in South Carolina and Georgia, but the use of the tides had been in effect before he was born. His inventions only replaced the crude machinery with which the planters had had to contend.

As late as 1830, and in some instances much later, the tides furnished power for small rice, grist, and saw mills in lower South Carolina, and in Charleston a saw mill was thus operated until 1871, as were several mills for various purposes in that city and in the coastal section of the state, until the Civil War.

When Lucas had built a number of pounding-mills, his son, Jonathan Lucas, Jr., was prevailed upon by the British government to go to England, where the milling methods invented by his father were attracting attention. There he built several mills for preparing rice for market, using the inventions of his father. Rice from other countries was being shipped to England to be prepared for market, and Lucas arranged to have the first shipment of rough rice exported from the province sent there. This opened the way for a large export trade. Our rough rice was soon being milled not only in England but in other European countries, and this went far toward establishing abroad the reputation of Carolina rice.

That the milling of Carolina rice in Europe must have made considerable progress is shown by the publication in this country, not many years ago, of a photograph of an old pounding-mill in Austria, above the door of which could be seen in large letters the words, "Carolina Rice Mill."

Early in the nineteenth century the steam-driven pounding-mills on the plantations were gradually abandoned, and the preparing of rice for market was transferred from the plantations to cities adjacent to the rice-growing section, where large mills were built and equipped to do much better work, both in hulling and in polishing the grain.

The largest of these mills and the last to be kept in operation were the West Point Mills and the Bennett's Rice Mill in Charleston. The original engines of the West Point Mills were imported from England about 1830, and were heavy, low-pressure, walking-beam engines, similar to those then in use in the side-wheel steamships. Thirty years later the low-pressure steam cylinders were replaced by high-pressure cylinders.

A few years ago, Mr. Henry Ford visited Charleston and was much interested in these old engines in the West Point Mills. The City of Charleston, having purchased the property upon which the mill stood, made him a present of the engines. I was told recently by the gentleman who now owns my old plantations, that he had presented to Mr. Ford the engine which had been used in my threshing mill for fifty years.

The happiest time in a rice planter's life was in the harvest season when his mill was steadily threshing his crop, the "beater" groaning and the engine laboring with its load. I remember well how I loved to watch the steady stream of rice pouring through the chutes into the measuring tubs and then being dumped into the "rice house," ready for shipment. The engine, which Mr. Ford now has, never failed me. Lucius Gadsden, my tall black engineer, delighted to brag of it. He ought to have been fond of it, for it never gave him the slightest trouble. In fact, it ran so smoothly that, lulled by the regular rhythm of its strokes, he dozed most of the day.

Lucius died before the engine, which he had watched over for so many years, was taken down and sent to Mr. Ford. Otherwise, I am sure that in some way he would have followed the engine to its destination, and would have tried to show how it should run. I have no doubt but that he would have been naïve enough to have offered both his services and those of the old engine to run the entire Ford plant.

My engine gave so little trouble that Lucius began to attribute its performance to his ability as an engineer. As a result he became much inflated with his own importance. On one occasion the captain of the schooner which carried my crop to Charleston, a great practical joker, persuaded Lucius that he was really too big for his job, and that he was thoroughly qualified for the position of engineer at the West Point Mills. Naturally this greatly appealed to Lucius—so much that when he had finished threshing my crop he took his wife and children and, full of great expectations, went to Charleston to apply for the job.

A month later, wet, bedraggled, and hungry, Lucius and his family walked all the way back to Combahee. His expectations had not been realized. A policeman, he said, had chased him off the wharf of the West Point Mills, when he had picked up a little wood to keep his family warm. He swore then and there that he would never leave Combahee again. He kept his word, and ran my old engine until he died.

4

THE PLANTING OF RICE

THE rice planters of my day occasionally discussed the question, Which counted the most in the making of a crop, the land or the man? And seldom did they agree. I used to maintain that during the early years of the industry, when the lands were naturally more fertile, and before the alluvial soil had begun to settle, and when the drainage for that reason was greater, the land counted more in the making of a crop than did the man who cultivated it. Men who were only fairly good planters often in those years made good crops.

In the later years of the industry it required more skill to make a high yield to the acre. The fertility of the soil, after years of planting, with little or no fertilizer, gradually lessened, and the level of the fields sank slightly from year to year. It has been estimated that through a period of a century and a half the rice fields of South Carolina and Georgia sank fully a foot, and perhaps more.

Another decided advantage the earlier planters had over those of later years was that the higher the surface of their rice fields, the less liability there was of breaks in their river banks, for the higher the land on the field side of the bank, the better able it was to withstand an abnormal pressure of the water, especially when storms occurred. This was a very decided advantage in Nathaniel Heyward's time. On several of his plantations, which I later planted, I am quite sure I had not only more, but larger and deeper breaks than had my great-grandfather during the fifty years he was in possession of these places.

The foregoing facts prove, I believe, that the land helped out the earlier planters more than it did those of my day, who had certain disadvantages to contend with which their forebears did not have. Hence, toward the end of the industry, in making good crops it was the man who should be given credit rather than the land.

The planting, cultivating, and harvesting of rice in the South Atlantic States differed greatly from that of other grain crops in the United States, and there are few now living who know anything of the methods used in the growing of a rice crop, and especially of the system of irrigation, so essential to its growth. This fact induces me to enter into considerable detail in describing the methods employed in growing rice in these states, with reference particularly to the years following the Civil War.

The first step in the growing of a rice crop was taken in the late fall or early winter after a killing frost. This was to burn the stubble of the preceding crop. Burning stubble was usually done by women, who dragged the fire with their hoes. When the stubble was thoroughly dry and a stiff breeze blowing, they sometimes had to jump across the quarter ditches to avoid the advancing fire. There was considerable excitement in this work, and the women seemed to enjoy it. Their dresses were tied up to their knees and did not hinder them from jumping the ditches when they were caught in a close place by the fire. "Look out, Sister!" they would often call to each other. "Don't let dat fire ketch 'ona. Jump across de ditch."

It was unfortunate for the good of the soil that the only way of getting rid of the rice stubble was to burn it, for it was quite thick, and by fall had grown to a considerable height. A story was told in Charleston of a gentleman who was a better merchant than he was a planter, and who owned a plantation which he seldom visited, conducting its operation from his office in the city. Having read agricultural articles advocating the enrichment of the soil by plowing under certain cover crops, he directed his overseer to plow under all the rice stubble. The overseer replied that this would be impossible; there was too much of it. The owner then asked why his overseer could not plow under only half of the stubble on every field, and the overseer said he did not know how to do that. "Why," said the merchant-planter, "that's very easy. All you have to do is to flow the field half-way up on the stubble, and then burn all of it above the surface of the water." The overseer assented, but never tried it, and later told the story on his employer.

When the stubble was burned the blackbirds descended on the fields in large flocks and picked up the unsprouted grains from the past year's crop. During the fall before, when the fields had been under water, the wild ducks had had their chance to feed on the shattered grains and often had come in flocks larger by far than did the blackbirds. Both ducks and blackbirds were thus helping the planters, for they were ridding his land of grains of "volunteer" rice which, when turned under by the plow, would have reduced the yield of his next crop.

Before 1850, few if any mules were used for plowing in the rice fields, which were kept under water until early in February, when the land was turned by Negroes with hoes. Each field hand was given a task, not more than a quarter of an acre, one-fifth of the amount plowed by a mule in later years. The surface of the tidal swamps is underlaid by a sour muck, and this muck, because of lack of drainage, is never improved. Its sourness was injurious to rice, and shallow planting was always necessary, whether the land was turned by hoes or plowed with mules. A clay topsoil could be plowed to a depth of two inches, but a dark peaty soil could not stand plowing deeper than an inch.

During the years of slavery, the Negroes, walking in the alleys of the previous year's crop, turned the soil with eight-inch hoes, set perfectly straight on the handles. This was known as "back sodding." Later, with hoes set at an angle, they chopped up the land exceedingly fine. In fact, they both plowed and harrowed it with their hoes, and from all I have been told they did most excellent work, often better than was later done with disc harrows.

When mules began to be used, rather crude wooden harrows with iron teeth, made on the plantation, took the place of the hoe, and in later years disc harrows were very generally used.

In my great-grandfather's time, after the fields had been turned with hoes, or, in my time, plowed, the next step was to clean the ditches, small as well as large. This work was done by both men and women, and the implements used were long-handled scoops, in which mud which had accumulated in the ditch was dragged out and scattered along its edge. Every year some of the ditches had to be sunk deeper. Only men could do this, walking in the ditches and throwing out the mud with shovels. In cleaning larger ditches and canals, scoops were used.

With us, rice has always been planted in trenches four inches wide and eleven inches apart and very shallow, only enough soil being displaced to cover the seed. From the beginning of the industry, these trenches also were made with hoes, and it is remarkable how straight the Negroes made them. In them the seed was sown by hand. Women always did this work, for the men used to say that this was "woman's wuck," and I do not recall seeing one of the men attempt it. My first two or three crops were planted by hand, and then drills began to be generally used.

There were two periods during which rice might be planted. The first was from the tenth of March to the tenth of April, and the second from the first to the tenth of June. Never was any planting done between the middle of April and the last of May. The reason for not planting after the tenth of April was the coming of the May birds, which, during the month of May, were on their way northward, after wintering in the far South, and could always be depended upon to appear in our rice fields. They seemed to travel on a regular schedule, and were always on time.

Early in May, a planter riding through his fields listened for the twitter of his little enemies, the advance guard of the flocks to follow, and seldom did he fail to hear it. If his crop was just appearing above the ground he knew that half of it was doomed.

Should the soil be damp and soft when the May birds came, by pulling at the rice appearing above the ground, they would also pull up the grain which they were seeking. If, however, the land was dry and hard, in trying to pull up the grain they would break off the blade above the ground, which meant the death of the grain when the "stretch water" was let into the field. The May birds would not remain in the rice fields longer than two weeks, but during that time they were capable of destroying a great deal of rice, and this could be prevented only by planting not later than the tenth of April.

By the tenth of September these same little birds would be back again. They had changed their plumage and were known as rice birds.[1] They soon fattened on the ripening rice and were always considered a great delicacy, but they were an expensive dish for our planters.

Rice planted in early June escaped both the May birds and the rice birds, but often did not yield as well as that planted in March or April; hence the percentage of the crop planted in June was usually not more than one-fifth of the total crop.

Some planters thought that by putting "bird-minders" in their fields and giving them muskets and plenty of powder, and keeping the birds moving by constant firing, they would decrease the damage, but I have observed that the rice birds'

---

[1] In Maryland and Delaware these birds are called reed birds, and in New England they are known as bobolinks. Since rice has ceased to be planted, they no longer stop in South Carolina on their way north or south, or at least relatively few do.

little craws were just as full when the sun set as they would have been had they not been molested.

There were other planters who never attempted to keep off either the May birds or the rice birds, claiming that the sound of guns attracted their attention and caused them to leave other plantations and congregate in the one where the guns were fired. This may have been true, for those little birds were pretty smart and may have surmised that there was a good reason for the firing of the muskets.

A lady from Virginia once asked me why I planted a crop which birds could destroy, and my only defense was that if she had ever seen rice growing, or had eaten rice birds, she would understand.

One of the principal reasons which prevented the planters of South Carolina and Georgia from competing in the production of rice with the planters of Louisiana, Texas, and Arkansas—the states to which the industry migrated—was our inability to use harvesting machines. Toward the end of the industry we had in use every machine capable of being worked in our fields. We used modern disc harrows and rice drills, and could have used tractors to some extent, but it was an impossibility for us to use improved harvesting machines. When a crop was ripe it had to be harvested as quickly as possible. Just as soon as the water left the surface of the field, the Negroes were in it with their rice hooks. We could not wait for the land to dry, for should a storm or a high wind occur and blow the rice down and tangle it up, it was impossible for either a Negro or a harvesting machine to cut and gather it. A machine could not stand up on the boggy soil, which had been under water for more than two months,

much less cut the rice. Hence, to gather their crops our planters were forced to have on their plantations a large number
of Negroes.

It was fortunate for the emancipated slaves that after
the Civil War they could not be replaced with machinery,
for many of them would have starved to death. Instead of
using machinery many of our planters, and especially my
father, thought they ought to give the Negroes enough work
to keep them as busy as they had been before they were
freed. As a result, I know that the few crops he planted cost
him far more than they should have; in fact, it was said of
him that he was "Negro poor," and there was much truth in
this.

5

IRRIGATING A CROP

BETWEEN the tenth and the fifteenth of March rice planters in South Carolina made, as I have stated, their first
planting, which was usually their largest. Their second was
made as early in April as possible. When the winter and
early spring had been favorable for preparing their land,
little or no rice was planted in June. Adjacent squares which
could be watered together were often included in the same
planting.

Early in the morning wagons loaded with sacks of seed
rice were on their way from the threshing mill to the field,
and throughout the squares to be planted that day sacks were
distributed so that little time would be lost in refilling the
drills. Planter, overseer, and foreman walked all day across
the "beds" to see that the drills were not skipping, while
close behind the drills followed the covering harrows. These
harrows covered the grains of rice very lightly, the more
lightly the better. A drill was set to sow two and a half bushels of seed per acre, and planted eight to ten acres a day.

When squares planted the same day could be watered
without interfering with those to be planted the next day,
before the overseer left the field he instructed the "water-
minder" to lower the inside doors of the river trunks, and

raise the outside ones so that the next flood tide would flow the field. Sometimes, depending upon the acreage to be flooded, it required only one high tide to cover the field completely, causing it to look like a large lake. This first flooding of the crop was known as the "sprout-flow."

A day or two after the field had been flowed, especially if there was any wind stirring, a considerable quantity of trash, consisting largely of rotten stubble and roots of the rice, would be blown against the check-banks, and hands were sent with long-handled rakes to gather it on the banks, where, when it had dried, it was burned.

The sprout-flow, as its name indicates, was used for a dual purpose: to hasten the sprouting of the seed and at the same time to kill any grasses which might be already sprouting, for neither grass nor rice, when in white sprout, can live under water.

The length of time the sprout-flow was kept on the field depended on the weather. When it was warm the seed sprouted quickly. Each morning the water-minder would grapple under the water for a few of the grains, and just as soon as they began to "pip," the inner doors of the trunks would be lifted, and usually it required only two ebb tides to drain the field.

It was necessary to dry the field as soon as the seed sprouted, for as long as water remained on it the rice would not throw a root, and within a few days the sprouts would become so stretched that they would slough off, and the vitality of the grain would become exhausted. In the meantime much of the grass in the white sprout had been killed.

When the weather was warm and the sun shone, the blades

of rice soon appeared above the ground. When they could be traced in long rows against the setting sun, the field was again flooded. This second irrigating was called the "stretch-flow," and to regulate this flow both experience and judgment were required on the part of the planter or his overseer, for this was the most critical stage of the crop.

Really the only object of the stretch-flow was to destroy grasses which had sprouted after the sprout-flow had been withdrawn. Again the water was to catch them in the white sprout. Had this not been necessary, better crops of rice could have been grown, for occasionally the stretch-flow was detrimental to the young rice, and not really essential to its growth.

To one riding through a rice field, the surface of the field appeared perfectly level, but when water was let into it this was seen not to be the case. The incoming water soon showed that certain places in the field were slightly higher and others slightly lower than the average level of the land. It was these high places and low ones, comprising altogether, in an average rice field, not more than 20 per cent of the entire acreage, which called for much skill to irrigate the rice during the stretch-flow. The weather and the condition of the rice also had to be carefully considered.

When the stretch-flow was let into the field, the water was at first kept for several days as deep as the check-banks would hold it. No land could be seen anywhere, and during this time the growth of the rice under water was carefully watched. When the rice on the higher places had grown to the length of one's forefinger, the water would be lowered to a depth which exposed the "hills," as we used to call them, as little as possible, and yet not so deep on the low places as to drown out the rice.

A field of rice, when the stretch-flow had been slacked down, was a unique sight. From 75 to 80 per cent of the crop would be standing straight up above the water. Ten per cent would be growing on dry land on high places, and another 10 per cent lying flat on the water in the lowest part of the field, the blades much stretched by their having endeavored, while under the water, to reach the surface and sunlight.

For a week, and sometimes a little longer, the stretch-flow was kept at the depth to which it had first been slacked down, and then for ten days it was gradually lowered until the field was entirely dry. This gave the rice in the low places, which had suffered from the depth of the water, time in which to throw a new leaf and become strong enough to stand upright.

The next stage a rice crop passed through was called the "dry growth," which usually lasted about forty days, the time depending on the rainfall. Should there be little or no rain, and should the crop begin to go backward, the time for putting on the "harvest-flow" would be advanced. This last irrigating was done solely to benefit the rice, for it was too late for the water to have any effect upon the grass, which only the hoe could then destroy.

While the fields were dry, between the stretch and the harvest water, they were laid out in half-acre sections, marked by stakes; and hoeing a half acre was a day's task for a full hand. Half hands were placed two in a half acre, and when it was possible their tasks adjoined those of their parents. The object of hoeing was not only to get rid of any grass growing in the alleys, but also to stir the ground which had become packed by the water.

Despite the bad work which the rice planters of my day had to accept, they continued during the dry growth to give their crops two hand hoeings, the last being shortly before the harvest-flow. By the time this flow was let into the field the rice should have tillered very considerably and been at least fifteen inches high.

For ten days or two weeks the depth of the harvest-flow was kept at the same gauge at which the water had been held when first slacked down during the stretch-flow, and from then on, every five or six days, it was gradually deepened. This continued until the rice was fully shot out, when as much water as the check-bank would hold was let into the field. This flooding was continued until the rice was ready to be gathered, though every ten days the water was changed.

I know of no crop which in beauty can be compared with a crop of rice. In my dreams I still see the crops I used to grow, and when I am awake, I am conscious of the fact that my dreams failed to do them justice. This was especially true during the late spring and summer months, when the crop was passing through its successive stages of growth and looked different to me each day as I rode through it.

Until the middle of July the color of the rice never looked the same. Some days it changed as constantly as the colors change on the surface of the sea. As its blades changed their direction with each shifting breeze, they changed their color also. Over the field a breeze often blew, coming inland from the ocean across the salt marshes and up the lower reaches of the river. Thus the crop was kept in constant motion, swaying in one direction and then in another. The result was that the whole field, as far as one could see, appeared to be alive, shifting with the wind, the sunshine, and the shadows of passing clouds.

As the season advanced, a decided change gradually took place in the color of the field, for its green began to be mingled with gold as the heads of rice appeared and its stalks began to be weighted down with the ripening grain. Yellow then predominated over the green until the whole field looked like a mass of gold, as it awaited the hook of the reaper.

Rice planted in March ripened the latter part of August, and, very shortly after, rice planted in April was ready for the sickle.

When the harvest-flow was let off the field, the Negroes would begin cutting the rice. They cut three rows at a time. Grasping the stalks with their left hands, they used the sickles with the right, laying the rice on the stubble behind them in order that the sun might dry it. The task of cutting rice was a half acre, and a good hand could do this in two hours. The next morning each hand would cut a quarter of an acre, and then wait until the rice cut the day before had been sufficiently dried to be tied in bundles and stacked in the field. Usually four or five stacks of rice were put in each half acre.

On some plantations these stacks remained in the field until they were dried out enough to be threshed; on others they were hauled to high land for safety. The latter was an extra expense, but I have known it to save crops. On my plantations, to move the rice to the barnyard was a long haul, and I always took the chance of leaving it in the field. In an experience of twenty-six years, I can recall losing only one crop by not removing it from the field. I counted on storms coming late in August or early in September, and when the tenth of the latter month had passed I felt my crop was fairly safe.

On every rice plantation there was a mill capable of threshing from six hundred to twelve hundred bushels each day. These mills were usually situated on a bluff on the edge of a river or creek, so that vessels could lie very close to the mill and the rice could be readily taken on board.

The mills consisted usually of three buildings: a "conveyor house," into which the bundles of rice were thrown from the wagons and carts hauling them from the field; the mill proper, a two-story building containing all the machinery; and a "rice house," where the rice was stored awaiting shipment. A "conveyor cloth" ran from the conveyor house to the second story of the mill, and on it the bundles of rice, after the bands had been cut, were carried to the "beater," which threshed off the grains. Within the beater box, iron rakes moved the straw to a chute through which it fell into the racks of wagons waiting to haul it away.

As the grains of rice were threshed from the straw, they were passed through a fan on the first floor of the mill, and then in elevator cups were carried again to the second floor and run through a large screen. Again falling to the first floor, they ran through a "market fan," after which elevator cups carried them back to the second floor and emptied them into wooden tubs, each usually holding fifty bushels.

A Negro boy sat on one or the other of these tubs, and as

soon as one tub was full he would turn the stream of rice into the other tub, and let the rice in the full tub run down a chute into the rice house. The hardest part of that boy's work was keeping tally of the tubs he dumped during the day. In my mill, old Abby Manigault was always in the rice house, also keeping tally, and if her tally and the boy's didn't agree, the next day there would be another boy on the job.

Seed for another year's crop was never threshed in the mill, for some of the grains would be cracked by the beater. It was threshed by hand on two-inch planks, and then run through the fans and the screen in the mill.

Fertilizers were not often used on a rice crop. Rice planters have never agreed as to their effectiveness, which irrigation may have lessened. Instead of buying fertilizers, some planters thought it paid better to rest certain squares which were failing to produce good yields.

The cost of growing rice on the plantations in South Carolina, after the slaves had been emancipated, was estimated, not including seed, at from twenty to twenty-five dollars per acre, this being the total expense of operating the plantation from the first of January, to the thirty-first of December. And our planters always hoped to make a yield of forty bushels per acre. Forty bushels of rice was to them what a bale of cotton to the acre is today to our cotton planters, and the rice planter made his forty bushels about as seldom as the cotton planter today makes his bale. For the last fifteen years that rice was planted on the Combahee, I would place the average yield at thirty-six bushels per acre, this not including years when crops were destroyed by storms.

As to the price of their rice the planters of my day counted on a d llar a bushel net on the plantation, which even before the year 1900 they were by no means sure of getting. When they received that price, their net profit was from twelve to fifteen dollars per acre. After 1900, with the exception of a few years, the price of rice gradually declined. I sold my first lot of rice in 1887 for five and three-quarter cents per pound, and it netted me, on the plantation, one dollar and forty-five cents per bushel. In later years I sold a lot which I had held for six months at fifty cents per bushel.[1]

After the Civil War and until 1890, the average prices of rice ranged between four and a half and five and a half cents per pound, and a good quality, weighing in the rough, forty-four to forty-five pounds per bushel, netted from a dollar and ten cents to a dollar and forty cents f. o. b. the plantation.

For many years rice was protected by a tariff, and the late Honorable John Randolph Tucker of Virginia, a member of Congress for a number of years, and one of the ablest and most charming men I have ever known, a staunch Democrat, and a believer in a tariff for revenue only, once told me that he had always favored a tariff on rice. At the time, I was a student at Washington and Lee University, and I

[1] For many years the price of Carolina rice was fairly stable. Through the kindness of a friend, Mr. James H. Fraser, of Georgetown, South Carolina, I have a copy of a chart of prices of rice taken from a publication entitled "Wholesale Commodity Prices at Charleston, S. C., 1790–1861," by George Rogers Taylor. The peak of high prices per pound shown in the chart was in the year 1817 when the price was seven cents. The next highest was six cents per pound in 1805, the year Nathaniel Heyward made a profit of $120,000, as stated in his life by his grandson, Dr. Gabriel Manigault. The third highest was in the year 1855, five cents per pound. The lowest price of all was in 1814, only two cents per pound.

have never forgotten the reason Mr. Tucker gave for taking the position he did in regard to rice. It was not, he said, because it was a Southern product, but because there was at that time as much rice grown in this country as there was foreign rice imported into it, and hence for every dollar of protection the planter got, the government received a dollar of revenue.

I knew nothing about planting rice at that time, but had fully made up my mind to engage in it just as soon as I left the University. Mr. Tucker's views on the tariff encouraged me greatly, and I often repeated what he told me, whenever an effort was made in Congress to reduce the tariff on rice.

6

A PIONEER PLANTER

OF THE men of my name who have planted rice in South Carolina, I have always felt that I had more in common with my great-great-grandfather, Daniel Heyward, though he died nearly a century before I was born. The bond which drew me to him was that he built the banks along the Combahee, which I used daily to ride. It was he who laid out and drained the fields along that river, which he was the first and I the last to plant.

In years gone by, as I rode through those fields, I would sometimes imagine I could see Daniel Heyward riding before me, wearing the red coat and black cocked hat which are seen in the old portrait of him, painted by Theus, which hangs in my dining room. As to that red coat, Daniel Heyward, most of his life and until South Carolina declared her independence from the Crown, was a loyal subject of the King, as were also the three men of his name before him in the province of Carolina.

I have always believed that I was destined to plant the fields which my family had planted. To grow rice must have been in my blood, for the only recollection I have of the old Physical Geography I was supposed to study at school was a picture of a scene on a South Carolina rice plantation, showing Negroes hoeing rice, and a planter in a broad-brimmed hat riding along a bank, watching them as they worked. Instead of studying my geography, I used to think that some day I would be riding through the fields I had heard so much about, and hoped that day would soon come. My hopes were fully realized.

When the first Heyward emigrated to Carolina is not known. It is certain, however, that his name was Daniel and that he came from Little Eaton, Derbyshire, England, where he had a brother, to whom he referred in his will. This will is recorded in the office of the South Carolina Historical Commission at Columbia and is dated September 14, 1684. He died shortly after it was made.

From the time of his coming to the province, Daniel Heyward made his home in Charles Town. Of his four children, only one, Thomas, survived him. This son also lived in Charles Town and during his short life must have attained to some local prominence, for he was appointed Powder Receiver, an important post under the proprietary government, and held that office until his death in his twenty-seventh year. He left one child, a son, who was born several months after his father's death and was named for his father.

This son, Thomas Heyward, the 2nd, was born in 1699 and made his home on James Island, which borders on Charleston Harbor. There he acquired a considerable amount of land, some of which he cultivated, but not in rice, for the island is completely surrounded by salt water. By his will it is shown that he owned some Negro slaves, but their number was small compared with those owned in later years by his son Daniel and his grandson Nathaniel. It was while he was quite a young man that there was an uprising of the Yamassee Indians, who had until then occupied lands to the west of the Combahee River. Thomas Heyward was captain of a militia company on James Island and took an active part in the campaign against the Indians. After the Indians were driven out, settlers were allowed grants in that section, long thereafter known as Indian Land, but now parts of Beaufort and Jasper counties. Thomas Heyward took up five hundred acres of that land. Possibly it was his intention to develop the property thus given him, but his death at the age of forty prevented his doing so. Had he lived longer, he, too, would probably have planted rice in the inland swamps of Granville County, and, if so, six generations of the family would have been rice planters.

Daniel Heyward, whom I imagined I could see riding through my rice fields, was the eldest son of Thomas Heyward, the 2nd, and was born July 20, 1720, on his father's plantation. Though he was destined to become one of the largest planters of his day, he was not among the first to plant rice in South Carolina.

The seed from Madagascar had been brought to Charles Town thirty-five years before his birth, and since that time had been grown on lands in the vicinity of Charles Town. Gradually the area in rice had been extended, but until the time of Daniel Heyward little reclamation had been done and scarcely any rice had been grown in the territory southwest of Charles Town, so recently taken from the Yamassee Indians.

Daniel was only nineteen years of age when his father died. As a boy, he must have rowed across the Ashley River

to attend school in Charles Town, and that he attended a
good grammar school is shown by the few of his letters
which have been preserved. Deprived himself of a college
education, he nevertheless fully appreciated its advantages,
for he sent his two eldest sons to England to complete their
education, and was only prevented from sending his two
younger sons by the Revolutionary War and his own death.

When he was about twenty-one years old he left the family homestead, and, in an open boat with a few Negro slaves
inherited from his father, made his way by an inland route
about seventy-five miles southwestward. The object of this
journey was to settle on the five-hundred-acre grant given
his father in Granville County. This land lay on Hazzard's
Creek, one of the tributaries of Broad River, a fairly large
salt-water stream, which empties into the Atlantic through
Port Royal Harbor. It was near Hazzard's Creek that
Daniel Heyward built his home, which he named "Old
House." There he resided until his death, and there he is
buried.

Little intention had Daniel Heyward, when he moved to
a new and sparsely settled territory, of confining his planting
operations to the five hundred acres given him by his father
or to the growing of only one crop. He soon began buying
other tracts of land and purchasing Negro slaves. The land
he could obtain at very low cost, and also the slaves, for the
demand for the latter at that time was not great.

In addition to rice, both indigo and cotton had been introduced into the province. The cry from the Mother Country
was for exports, and the successful production of these products looked exceedingly hopeful. All three of these Daniel

Heyward had in mind when he bought land and Negroes, for the country in which he had located seemed admirably adapted to their growth. There were swamps through his land which could be drained and where rice could be irrigated. In the low lands, indigo could be planted, and his high lands, on account of their fertility, were well adapted to the cultivation of cotton. These crops he planted for some years, but he gradually decreased the acreage devoted to indigo and cotton, while greatly increasing his acreage in rice.

The dwelling at Old House faced south and stood on a low bluff. Before it lay a great expanse of salt-water marshes, through which, not far from the house, flowed Hazzard's Creek. Looking across the marshes one could see distant woods, and to the southeast a wide opening on the horizon. Behind the house to the north, there extended for miles great forests of long leaf pine, untouched by the woodman's ax, forests unsurpassed anywhere in the South.

A winding avenue of very old live oaks today leads from the place where the dwelling house once stood to the public road. No care has been taken of these old trees; a few are dead and others are slowly dying. Nor is there anything to indicate the location of the house, not a single brick from one of its old chimneys, nothing except an open space in which no trees have grown. There is, however, no difficulty in locating the spot where the pounding-mill stood, for a short distance to the right lie the two large mill stones brought from Scotland, which once were turned by the power of the tides.

To the left of where the dwelling stood, and just beyond what must have been a flower garden, is the family burying ground, its brick enclosure still intact. There Daniel Heyward and some of his immediate family are buried, among them his eldest son, Thomas Heyward, Jun. (for he so styled himself), one of the four South Carolina signers of the Declaration of Independence, above whose grave the General Assembly of his state has in recent years erected a handsome bronze bust.

The grave of Thomas Heyward, Jun., lies close beside that of his father, and this, I am quite sure, was the wish of both. Daniel Heyward, from all I can gather, was very fond of his distinguished son and bequeathed to him the best of his lands on the Combahee River. These were some of the lands he had gotten through grants, during the latter part of his life, and he was greatly interested in the reclamation of their large tidal swamps. He was an expert judge of land. This is fully proved by the fact that the lands he selected for cultivation remained in the possession of his family and were planted in rice for one hundred and thirty-seven years.

There is an old family tradition that when Thomas Heyward, Jun., was doing all in his power to have South Carolina declare its independence of England, and was hoping that his name might be affixed to such a declaration, his father told him if it was so affixed he would probably be hanged for it.

From his boyhood, Daniel Heyward had a most kindly feeling for the country from which his people came. I expect, however, that on one occasion this kindly feeling of his ran pretty low; for riding one day in the course of the Revolutionary War to one of his plantations, he found it being pillaged by several British soldiers. Promptly he armed himself and two of his overseers and fired on the soldiers, who retired but soon returned with their number considerably increased, and Daniel Heyward and his men had to beat a hasty retreat.[1]

Of Thomas Heyward, Jun., statesman, soldier, and patriot, his father could well have been proud. Although he is not a direct ancestor of mine, being a half-brother of my great-grandfather Nathaniel, I have always wished I could claim a closer relationship.

White Hall, the country home of the Signer, was near Old House, both facing on the same creek and overlooking the same expanse of marshes. The former, however, was a much more pretentious place. Judging by the brick foundations which remain, the house must have been large for the times during which it was built. Behind it are four rows of live oaks, through the center of which leads a wide driveway, extending for nearly a quarter of a mile to the "big road." No finer or handsomer live oaks are to be found anywhere in the South, and I wish Thomas Heyward, Jun., could see them now, for they were only small trees when he lived there. How few realize when they drive beneath these trees, that the man who years ago planted them never saw them as we see them today. He planted not for the present, but for the future, not for himself, but for posterity, for the live oak is a slow-growing tree.

When President Washington made his visit to the South on his way between Charleston and Savannah, he spent the night with Thomas Heyward, Jun., at White Hall. The only buildings now standing at White Hall, which were there at

[1] From an old letter in the possession of one of the Heyward family.

the time of Washington's visit, are the quarters of the house servants, two large, white two-story buildings, constructed of oyster shells and lime, a composition known as tabby, which have stood fairly well the ravages of time. I am quite sure that Washington was surprised to see such large and comfortable quarters for slaves, and in such a sparsely populated country. He had seen few like them in Virginia or on his way south.

Both Old House and White Hall are located only a very short distance from Lemon, or Daws Island, one of the historic places in America, for it was on this island that Jean Ribault landed in 1562 and claimed the country in the name of the King of France. Upon a hillock on the island, he erected a stone pillar engraved with French armorials.[1]

For two centuries the people of Beaufort and other citizens of South Carolina hunted for that stone and failed to find it. It was never known what had become of it until 1905, when Woodbury Lowery published his *Spanish Settlements within the Present Limits of the United States,* and disclosed the fact that in 1564 a Spanish captain pulled up the stone and took it to Spain. A more recent investigation discovered a letter, written some months later, announcing the arrival of the stone in Seville.

Not long ago I visited White Hall and walked through those old slave quarters. I searched the walls for any writing, for the names of those who had once occupied them, and then I suddenly remembered that slaves could not write. They could leave no signs behind them. Their names had died with them long ago.

[1] Edward McCrady, *The History of South Carolina under the Proprietary Government* (New York, 1897), p. 45.

7

THE INTRODUCTION OF SLAVERY

SLAVES were brought to America in large numbers from different African tribes. The slave traders cruised for hundreds of miles up and down the west coast of Africa seeking cargoes where they could purchase them to the best advantage. Little did it concern them where the slaves they bought came from or to what tribes they belonged. What concerned them most was whether the slaves looked healthy and fit to work, and also where they could be disposed of most readily. Some of the Negroes were sold in one port and some in another.

It was not long after the discovery of America that the Portuguese and Spanish traders brought human cargoes from the west coast of Africa to Brazil and the Caribbean islands. The Spaniards began the trade for the purpose of developing their recently acquired possessions in the New World, and the English, for the same reason, were soon induced to engage in and to encourage the selling of slaves to their colonies in North America. Last to take part in this trade, but by no means the least active after they began, were the slave traders from the coasts of Massachusetts and Rhode Island. It was in their ships that most of the slaves sold in the South were brought.

It is not definitely known from which part of Africa the Gullah Negroes, who lived on our South Carolina and Georgia plantations, were brought. The supposition is, however, that they were among the Liberian group of tribes, and for this there appears some foundation. In a copy of the *National Geographic Magazine,* published in 1922, there are pictures of numbers of African tribes, and those of Liberian Negroes remind me, more than the others, of the Negroes I knew on the Combahee. I am quite sure that far more of these Negroes than those of other tribes were brought to South Carolina and Georgia, and their peculiar language, a mixture of African and English words, became more generally used in that entire section. This continues, although to a lessening extent, even to the present.

I also believe that our Gullah Negroes today are of purer African blood than any other Negroes in the world, with the obvious exception of those on the Dark Continent. Among them, even now, mulattoes are rare exceptions. Not only is their blood relationship closer to the present natives of Africa, but in their way of thinking and expressing their thoughts they more nearly resemble native Africans than do any other Negroes in the western hemisphere. Today I never see a motion picture of Africans, taken in their native jungles, with their earrings and nose-rings, without noting among them strong facial resemblances to certain Negroes I used to know.

On the best authority now obtainable, Daniel Heyward owned, at the time of his death, one thousand slaves, many of whom he purchased during his latter years, when he began the reclamation of his lands on the Combahee River. In such an undertaking he certainly needed them, for there was at that time in the province no white labor which could perform the work of reclaiming the river swamps. The white man could not stand the summer heat, nor could he endure working in the water. Negroes alone had to be relied upon.

Despite the fact that nearly all of the Negroes who once worked for me were direct descendants of the slaves whose labor, years ago, reclaimed the fields I planted, I do not recall ever hearing these Negroes refer to them. The work had been done; the slaves who did it had died too long ago, and they knew nothing about them. I can myself, however, recall very distinctly an old woman who came within ten years of having been born a slave of my great-great-grandfather.

During the spring of 1887, I visited for the first time since my boyhood the plantations I had inherited, and rode through the "street"[1] of the Negro settlement on Amsterdam plantation with old Squire E. T. Jones, who had been my grandfather's overseer from 1855 to 1866, and then for twenty-two years had continued to manage the Negroes of this plantation. Mr. Jones wanted me to speak to an old woman named Affy, who, he felt quite sure, was the oldest Negro on the Combahee.

To this day I can recall Affy very distinctly. She was quite spry and active, and did not appear to me to be as old as Mr. Jones thought she was. She seemed very glad to see me, saying she had not seen me since I was a small boy. She talked to me of my grandfather and of my great-grandfather, neither of whom I had known. Later, the next fall when I

[1] Negroes on rice plantations always called the open space between the rows of houses in their settlements "the street."

returned to the plantation, I learned from Mr. Jones that Affy was dead. We looked at the record my grandfather had made of his "emancipated slaves," dated July, 1865, a record which in later years I referred to on many occasions when Negroes who had belonged to him would often come to me to tell them their ages, always calling the record "de book." In my grandfather's writing I found this:

Name	Age	Occupation	Residence
Affy	78	Midwife	Amsterdam

Just to the right of Affy's name, and in my handwriting, I see in the record which now lies before me, "Died October 1887, 100 years old."

Affy certainly had a most remarkable life. She was born the slave of Thomas Heyward, Jun., and all of her life lived in the same house. From his possession, she passed, with the plantation, into that of his younger brother Nathaniel, and at the time of his death she was sixty-four years old. She then became the property of his son Charles, my grandfather, remaining his slave for fourteen years, until at the age of seventy-eight she was emancipated. After this, as a free Negro, she continued to live in the settlement until she reached the century mark. Therefore, if my great-great-grandfather had lived ten years longer, I should have had the unique experience of having known one of his slaves.

It was indeed fortunate for Daniel Heyward that when he moved from his father's home on James Island opposite Charles Town to begin his life work in the wilds of Granville County, he was able to take with him several slaves who had been raised and trained by his father, for these Negroes could understand what was said to them, and also could make themselves understood by those who were accustomed to them. They had probably been in this country for one or more generations, living in contact with white people, and had learned to talk the Gullah dialect.

Having been thrown much of my life with Gullah Negroes and knowing their natures fairly well, I have wondered if they had not inherited from their ancestors a trait often noticed—that of pretending to misunderstand what was said to them when it suited their purpose to do so. They sometimes made me think of a Negro at whose house a traveler knocked one night, to see if he could get someone to row him several miles up the river against a strong tide. Arousing the Negro from his slumbers, the traveler asked him if he could row. He replied that he could not. The traveler hired him to go along anyway, thinking he might be useful otherwise. Taking the oars himself, he had a hard and tiresome pull before reaching his destination. As he rowed, it provoked him more and more to see how comfortable the Negro looked sitting in the stern of the boat, and when the landing was finally reached he said to him: "Why in the devil, living right on a river, can't you row a boat?" "I kin row boat, Boss," the Negro replied. "When you come to my house an' call an' ax me if I could row, I t'ought you mean if I could roar like a lion, an' I tell you 'No, Sir.' Me no know it was row boat you mean."

It may be a hard thing to say, but it is not so meant, that the slaves who were brought in the slave ships were wild African savages. They differed in no degree from some of those who today remain in the jungles of their native land. That the first generation of slaves brought to Carolina should have shown little mental improvement during their lifetime is not to be wondered at, for how could they? They had been transferred from a jungle in Africa to a wilderness in the New World, where they associated only among themselves, scarcely ever coming in contact with white people, whose language they could not understand, nor could the whites understand a word they said. I am quite sure that these slaves, as long as they lived, with few exceptions, learned the use of very few English words; the few exceptions among them being some who were especially bright or those who were children when they were brought over.

To show that this was the case, I would refer to the last shipments of slaves brought to the United States long after the traffic in human beings had been forbidden by the Federal government and by nearly all of the Southern states. Few today know of the slave ship "Wanderer," whose owners, shortly before the Civil War, attempted, in defiance of the law, to bring a cargo of slaves into Georgia and South Carolina.

The "Wanderer" was captured and her cargo of Negroes sent back to Africa. Accompanying the "Wanderer" on her voyage to the South were two other ships, both with cargoes of slaves, and of these ships little has ever been known. For a while, dodging among the sea islands on the Georgia coast, they managed to evade the Federal gunboats, but, realizing they could not escape capture, one of the vessels, on a dark night, ran up the Savannah River and the slaves were hidden on one or more of the islands which lie in that river some miles above the city of Savannah. There a number of these Negroes remained as long as they lived, many of them living to be quite old. They built huts and grew crops, and when a freshet occurred they sought safety on the highlands on the Georgia side.

Their owners tried to sell some of these slaves, but found it a difficult matter, for those who might have been interested in buying them for a low price, were beginning to realize that the continuance of slavery was uncertain. Also, the planters of that time had become accustomed to working Negroes born and bred in this country, who understood what was said to them and did not have to be followed foot to foot. Some of the slaves were taken through the middle section of South Carolina, and a few sold in Richland County, where Columbia, the capital of the state, is located, and where, I have been reliably told, the Negroes as well as the white people were afraid of them. A member of a prominent Richland County family bought several of these Africans, one of whom, a man, remained on the plantation until his death, only about twenty years ago. This old Negro was known by the name of Joe Adams, and the gentleman who now owns the plantation where he lived tells me that he was of quite a low order of intelligence. It was most difficult, if not impossible, to understand what he said, and his front teeth were filed quite sharp.

The second generation of Negroes left on the islands in the Savannah River were a great improvement on the first, and with the third generation the improvement was even more marked. Some years ago, curiosity prompted me to have an investigation made to ascertain what had become of these

Negroes, and I found, as I expected, that all who had come from Africa had died, and their descendants had moved away from the islands and had become so entirely amalgamated with others of their race in that part of the country that they showed no difference in appearance, habits, or intelligence.

In this connection I might say that I have always wanted to know which have prospered more: the descendants of the Negro slaves who eighty years ago were the last to be brought to this country, or the Negroes, who, born and bred in the United States, sailed in 1878 on the bark "Arzor" from the port of Charleston, to make their homes in Liberia.

There could not possibly have been any difference intellectually, morally, or in any other respect, between the slaves first brought to this country and those smuggled in a century and a half later. A Combahee Negro of my day would have certainly pronounced them "one one."

On the other hand, there can be no question but that after the slaves were brought to this country each succeeding generation advanced mentally and that this advance has continued to the present day.

A very large percentage of the slaves owned by Daniel Heyward were born in Africa, and purchased in the slave-market of Charles Town. I can well imagine he had a hard time managing those untamed and untrained Africans, and yet I must believe that he was interested in their welfare and took care of them. Of this I shall cite one instance.

It had been the custom of rice planters in Carolina, during the days of slavery, to import from England the cloth with which they clothed their slaves and also to import for them blankets and shoes. During the Revolutionary War this became impossible, but it did not deter Daniel Heyward. He proceeded to build on his plantation a small cotton factory to manufacture cloth for his Negroes. This little factory was one of the very first to be built in America, certainly the first in the South, for it was operating in 1777. In a letter to his son Thomas Heyward, dated July 19, 1777, in referring to this factory of his, he said, "My manufactury goes on bravely but fear the want of cards will put a stop to it, as they are not to be got: if they were there is no doubt but we could make 6,000 yards of cloth in the year from the time we begun."

In this little manufactory he must have worked slaves. With the Revolution in progress, it was impossible for him to have employed white labor, even if there had been any in the neighborhood of his plantations. The fact that he had among his slaves some who were capable of weaving cloth clearly shows the advance they had made in intelligence since their forebears had been brought to the province.

What I have said in this chapter as to the degree of savagery which characterized the slaves brought from Africa is not intended as a reflection on our Negroes of today. They are not responsible for their antecedents, and they deserve great credit for the advancement which they, as a race, have made since slavery was abolished. And it might be added that nowhere does history show that Africans, deported from their native land, or emigrating of their own accord, have ever made the same progress, intellectually or morally, as have the descendants of the slaves of the South.

8

THE LARGEST PLANTER OF HIS DAY

IN THE city of Charleston is a portrait of a very old man, sitting in an arm-chair, with his right hand resting on a map of several rice plantations. Through an open window beside him is a scene on one of these plantations, with its growing crop, and in the distance is a threshing mill, standing near the river bank. Thus is most fittingly portrayed Nathaniel Heyward, the greatest rice planter of his day. His face, like that of his father, is clean-shaven, but sterner, and his eyes, undimmed by age, seem to look through one.

The picture was painted shortly before his death, in his eighty-sixth year. Sixty-odd years of his life were spent as a planter on the Combahee; and within a few yards of the spot where for fifty years he had his residence on a high bluff overlooking the river his solitary grave can be seen.

Nathaniel, the second son of Daniel Heyward, by his second wife, Elizabeth Gignilliat, was born in Charles Town, South Carolina, January 18, 1766. His birth there must have been due to the fact that his mother went to the city on account of the event, for at no time during his life did Daniel Heyward live in Charles Town. When he settled at Old House, he made it his permanent home, living there both summer and winter, as was the custom of practically all the rice planters of his day.

It was some years before the first settlers in the province began to realize that the white man could not withstand the malarial fevers which began gradually to increase and which they believed were caused by either bad air or bad water, a theory which was held until recent years. Little did they realize that the slaves who were being brought from Africa, though to some extent immune to malaria, might be bringing with them in their blood the malarial germ, which found ready carriers in a certain species of mosquito infesting the swamps and singing from the setting to the rising of the sun. This theory—that malaria was brought to lower South Carolina by the slaves—is now, I understand, accepted by certain medical authorities.

The boyhood of Nathaniel was spent at Old House, and until he was ten years old he had not visited Charles Town. Long before this, his brother Thomas had moved there, his house being on Church Street,[1] as had also his brother William, who built a house on Legare Street. This home of William Heyward's, which is still standing, was one of the first residences to be built on that old street. It is now owned by the Smythe family.

It was during a visit to his brother William's home that the first important event occurred in Nathaniel Heyward's life, one to which he occasionally referred in his later years. From the roof of his brother's house, on June 28, 1776, he

[1] The home of Thomas Heyward on Church Street is now owned by the Society for the Preservation of Old Dwellings, and is in excellent repair. It was there that President Washington stayed during his visit to Charles Town.

witnessed the battle between the British fleet under Sir Peter Parker, and the Americans holding Fort Moultrie on Sullivan's Island. The sight of this spirited engagement made an impression on the mind of the young boy which he never forgot.

When he was fifteen years old, he again went to Charles Town, where he must have remained some time, for while there he served with the Charles Town Battalion of Artillery, composed of South Carolina Militia, and commanded by his brother, Thomas Heyward, Jun.

In his book entitled *Traditions and Reminiscences of the American Revolution*, Joseph Johnson, M.D., of Charleston, says in a footnote: "My father often spoke of the late Mr. Nathaniel Heyward, then a stripling, assisting at this Station, in all the light duties of his brother's company, in patrol and sentinel duty, etc. Mr. Heyward not only confirms this, but remembers handing cartridges to my father while working the cannon. Mr. Heyward lived until April, 1851, probably the only survivor, previous to that date, of all who bore arms in the siege of Charles Town."

When the city was taken by the British, Nathaniel returned to his home to be with his family, his father having died nearly three years before. Until he reached manhood, his time was spent between Old House and Charles Town. Shortly after attaining his majority, he went with one of his cousins on a visit to Europe, most of their time being spent in England and France.

While in the latter country, they visited Versailles. There on a Sunday, when they were in the corridor of the palace, they saw Louis XVI and his Queen, Marie Antoinette, on their way to the Chapel.

It may seem strange, but this visit to Europe did not seem to have made any great impression on Nathaniel Heyward, for he had had very little opportunity to acquire a good education, and his reading of history had been limited. Though he was in Europe, his heart was in America, and he was glad when the time came to return to his familiar rice fields, where he could devote his time and energy to the occupation he had decided to make his life work.

Upon his return home, he took charge of the property left him under his father's will. He was disappointed when he realized that he and a younger brother had been left the smallest shares. Daniel Heyward, a believer in the old English law of primogeniture, had given to his eldest sons his best plantations. Nathaniel was left only two small plantations, one an inland swamp place not a great distance from Old House, and the other, consisting of some back fields only partly reclaimed, near the Combahee River and draining into it. In the inland swamp he planted his first crop, and lost it. Realizing what he had to contend with, he moved over to the Combahee and began work on his little place.

Shortly after this he married Henrietta Manigault, the daughter of Peter Manigault, speaker of the House of Assembly, an ardent patriot and a warm advocate of the Revolution. His father, Gabriel Manigault, had been considered the richest man in the province, a large merchant and planter, and a heavy contributor towards the independence of the province.

It was indeed a strange coincidence that Nathaniel Heyward, who later became such a large slaveholder, should have married the granddaughter of Gabriel Manigault, who had been strongly opposed to the continuation of the slave trade. He did not advocate the emancipation of the slaves already in the province, fearing it would be an unwise step for the reason that these Negroes were by no means, at that time, civilized enough for such a change, but he did not want to see any more Africans brought to our shores. Gabriel Manigault looked into the future and saw aright.

Nathaniel had been on the Combahee only a few years when his brothers, Thomas and James, employed him to take charge of their plantations. Thomas, more of a states-man and lawyer than a planter, was practicing law in Charles-ton, and James was spending most of his time in Europe and Philadelphia. In the management of these properties, by close application to his business, Nathaniel soon demonstrated his ability as a planter. He was too full of energy and con-fidence in himself to be satisfied to work for others and to plant only a small plantation of his own. It was fortunate for him at that time that his wife, an heiress in her own right, was able to contribute to his rice-planting industry the sum of fifty thousand dollars, for it at least gave him a start in carrying out his ambition to become a large planter. From then on, he began gradually to buy one plantation after another on the Combahee, and with the plantations he pur-chased the slaves who belonged on them.

The first plantations he bought belonged to the Gibbes, an old family in the province, who were not successful as rice planters. The Gibbes holdings on the Combahee con-sisted of several fine plantations known as Lewisburg, the

Bluff, and Rose Hill, the latter being practically the same plantation as Pleasant Hill. All of these places lay along the Combahee, at the best location on the river for planting rice, at the right pitch of tide, and at points where they were least exposed to freshets or the encroachment of salt water. Nathaniel Heyward was fortunate in being able to buy these places, for two of them lay in the same river swamp as did the plantations of his brothers.

Shortly after his purchase of the Gibbes property, he bought from his brother Thomas the best lands he had inherited from his father, nearly all of them grants, but it was many years before the property inherited by his brother James came into his possession through the widow of James dying without an heir. Owning the former Gibbes plantations and those of Thomas Heyward, he soon began to prosper, but never did he let pass an opportunity of buying a rice plantation on either side of the Combahee.

There is a tradition in his family that Nathaniel Heyward, while in Europe, visited Holland and was much impressed by the system of irrigation which he saw practiced there, and that upon his return he gave Dutch names to several of the plantations he subsequently owned. There was Amsterdam and Rotterdam, Hamburg and Copenhagen. But it was one thing for "Ole Maussuh," as the Negroes always called him, to give his plantations such names, and quite another to get his Negroes to call them by these names. Amsterdam and Rotterdam were too much for the slaves, and of their own accord they changed them to "De Swamp" and "De Lower Swamp," and these names stuck in spite of "Ole Maussuh."

On the Bluff plantation, when Nathaniel Heyward purchased it, there had been built a comfortable house, and there he lived until his death. The plantation undoubtedly got its name from a high bluff which overlooks the river, and a finer location could not have been found in South Carolina. The highland of the Bluff, which extended to the river, had rice fields on both sides: Rose Hill on the west, separated only by a canal, and on the east, all the lands which Nathaniel had bought or inherited and which lay between the Combahee River and Cuckols Creek, a bold stream emptying into it.

The house on the Bluff was a fairly large one, well built, but by no means an imposing mansion. It was typical of the homes of most of the rice planters, not only on the Combahee, but throughout the rice section of the state. That these homes were not finer than they were was due to the fact that on account of having to leave their plantations during the summer months the planters were compelled to have two homes, one in the country and the other in the city. The only exception to this was in the neighborhood of Georgetown, where there were a number of wealthy planters whose plantations lay near the seashore, and they needed to have only a cottage on the beach, in which they could sleep at night during the summer, and hence they could well afford to build fine residences on their plantations. A few of these old houses still remain.

Some of the finest houses to be seen in Charleston were built by rice planters, of lumber brought from their plantations; and the workmen who built them were slaves. Charleston for a long time was the favorite summer resort of rice planters whose plantations lay within sixty or seventy miles of the city. In this connection, a story is told on Nathaniel Heyward, for the truth of which I cannot vouch.

On one occasion, it is said, a committee of business men called on him and asked that he subscribe to the stock of an industrial corporation which they were endeavoring to promote. To their surprise Mr. Heyward refused. Noting their surprise, he said his refusal was due to the fact that "such industrial enterprises would ruin Charleston for what it was intended to be, a summer home for rice planters."

Today, with one exception, none of the homes built on the Combahee before the Civil War remains. They were all burned the last winter of the war when General Sherman's army passed through South Carolina, and its stragglers wandered over the deserted plantations.

Except for an open space close to the live oaks at the Bluff, one could not now tell where Nathaniel Heyward's home stood. Nor is there any trace left of his summer home in Charleston, which stood on the corner of East Bay and Society streets, facing the Cooper River, and immediately opposite Bennett's Rice Mill. This house was a large one, with columns the height of two stories in front, and with the main porch on the side facing south, as do many of the old houses of Charleston. In front, a high and handsome gateway, with four large brick posts, opened on East Bay.

I am quite sure that when he sat on his front porch, from which there was a splendid view of Charleston's beautiful harbor, with Fort Sumter in the distance, Nathaniel Heyward's gaze often wandered to Bennett's Rice Mill, as he thought of the rice crop ripening in his fields, which would soon be brought to Charleston to be milled. I wonder if he figured on paper the number of bushels he would ship—a bad practice often indulged in by rice planters, as I have found out by sad experience.

On the Combahee, they often told this story on Nathaniel Heyward, for which there may have been some foundation. During the winter of 1846, when he was eighty years old, he was taken quite ill while on the Bluff, and his physician, Dr. Thomas L. Ogier, came to Combahee in a small steamboat to carry him to his home in Charleston. The boat reached the plantation late at night, and the Negroes in the settlement knew nothing of its arrival. Next morning it started back down the river, and the Negroes, who had never before seen a steamboat, did not know what to make of it. They became excited and frightened. A large number of them were so worked up that running along the river bank they followed the steamboat down the river, as far as they could, hollering at the top of their voices, "Oh, me Gawd! Oh, me Gawd! De devil got Ole Maussuh!" I wonder if he heard them.

I recall Dr. Ogier quite well. He lived to be ninety years old and attended five generations of my family, from Nathaniel Heyward to my children. I wish now that I had asked him about this trip of his to Combahee.

All of his life Nathaniel Heyward kept in close touch with everything being done on all his plantations, and how he managed to do this without a clerk or stenographer is now hard to understand. Even his correspondence with his factors in Charleston must have been considerable. The only explanation is that he employed competent overseers. The majority of these men were young Englishmen who had come to make their homes in South Carolina.

On the night of April 9, 1851, Nathaniel Heyward went to bed as well as usual at his accustomed hour. The next morning a young Negro girl, named Mary, went to his room to kindle a fire for him to dress by. Returning in half an hour, she found him still in bed. She called him, and receiving no response went to his bed and found that he had passed away. Often did Mary, who later became the wife of Lucius, my engineer, tell me of that most momentous morning of her life.

A great-grandson of Nathaniel Heyward, when writing of his death, quoted these appropriate lines:

> Time had only laid his hand
> Upon his heart quietly, not smiting it,
> But as a harper lays his open palm
> Upon his harp, to deaden its vibrations.

Long before he died, he had selected a spot in his rose garden where he wished to be buried. It is a short distance from where his residence on the Bluff plantation once stood, and only a few yards from the Combahee River. There Nathaniel Heyward, the great slaveholder, was placed beneath the soil he loved so well, while around his grave stood one thousand of his slaves who sang, in their Gullah dialect, one of their own hymns.

On the tablet above his grave is engraved the following:

Nathaniel Heyward
Who departed this life April 10, 1851
aged eighty-five years.

While yet a youth, he shared in the vicissitudes and privations of the Revolution, and having thus early imbibed the principles of constitutional liberty, he undeviatingly adhered to them.

Planting on an extensive scale for more than 60 years, he illustrated by his prosperity and success and the attainment of a green and honorable old age, the truth of the classic maxim.

> "Nihil est agricultura melius,
> Nihil uberus, nihil dulcius,
> Nihil homine libero dignius."

His firmness and even stoicism as a man were tempered by his urbanity and liberality as a gentleman; and he fulfilled with marvelous dignity and wisdom the various duties of life.

A residence here of half a century with the association and attachment induced him to select this spot as a resting place for his remains.

9

THE INCREASE OF THE SLAVES

THE Negro settlement on the Bluff plantation was a short distance from the dwelling house, and today, although it contains fewer houses, it looks very much as it must have done when it was first built.

The houses in this settlement differed in one respect from those in most Negro settlements in the former rice section of South Carolina, in that they were double houses in which two families lived. There were four rooms, divided through the middle by a partition, and each section had a front and a back door, and three windows. There was a brick chimney on each end of the house, with a large open fireplace. They must have been built of excellent long leaf pine lumber, for now, though much over a hundred years old, they are in fairly good condition.

In the Bluff settlement, as in all others of its kind, the houses were placed in two long rows, facing each other, with a wide space between the rows. There the Negroes made their favorite rendezvous, when not at work, and there they talked to their heart's content. Old men were seldom seen in these gatherings; they much preferred their own doorsteps, where they sat all day smoking their pipes, which often went out as they dozed in the sunshine. Usually the old men outnumbered the old women on the doorsteps, for the latter, unless very feeble, preferred to spend their time fishing in the trunk docks, beside which they could sit the entire day, silently and patiently waiting for a bite.

At the time of Nathaniel Heyward's death, there were ninety-eight slaves of all ages living in the Bluff settlement. This number did not include house servants, who had their own quarters quite close to his house and associated little with the other Negroes. An inventory of his personal estate shows that he owned two thousand slaves. Some years previously, he had given to his son Arthur one hundred for his Ogeechee plantation in Georgia, and at least two hundred to each of his married daughters for their husbands' plantations on the Cooper, the Pon Pon, and the Savannah rivers. Hence at one time he must have owned at least twenty-five hundred slaves.

I have often been asked if I thought he knew them all and have said I did not see how he possibly could have done so. They were scattered over seventeen plantations and seldom did he come in contact with them. That he had all of their names listed I am quite sure. These, of course, were given to him by his overseers.

There are no records which show from what sources Nathaniel Heyward acquired such a large number of slaves. He inherited a few from his father, and there were quite a number already on the plantations he purchased from time to time. To these may be added a small natural increase during his long life as a planter, but the large percentage of his slaves must have been bought in Charleston during his first ten years as a planter when he was carrying on a large amount of reclamation work. For the slaves he bought in Charleston, he paid high prices, because during the Revolution the losses in slaves, from various causes brought about by the war, were large. As a result, the demand for them greatly increased, not only for reclaiming tidal swamps for the culture of rice, but also for the growing of cotton and other crops throughout the state.

It would be interesting to know the rate of the natural increase of Nathaniel Heyward's slaves during the sixty years he planted rice. Of one thing I am quite sure—he did not enjoy the same good fortune as did his wife's grandfather, Gabriel Manigault, in having so many of the offspring of his slaves grow to manhood and womanhood. My reason for saying this is that in 1790, "when an examination before a committee, appointed by the House of Commons in England to inquire into the treatment of slaves in the colonies then held by Great Britain" was being conducted, it was brought out in evidence that in its former Colony of South Carolina a part of the slaves of Gabriel Manigault had in thirty-eight years increased from eighty-six to two hundred and thirty, without any purchase of slaves having been made except to replace twelve or fourteen old slaves with the same number of young ones.

It is not surprising that with such a high rate of increase among his slaves Gabriel Manigault, my remote ancestor, could have well afforded to oppose the bringing of more slaves into the province, and also to have contributed seven hundred pounds to the College of Philadelphia, now the University of Pennsylvania.

I have every reason to believe, from what has been said

to me by former slaves of Nathaniel Heyward, that he must have taken good care of them. I am quite sure, however, that the natural increase among his slaves could not have compared with the increase said to have been among those of Gabriel Manigault. Some mistake must have been made in the evidence before the British House of Commons.

It is a well known fact that on the rice plantations of South Carolina and Georgia there was a very high mortality rate among the infants of the slaves during the time of Nathaniel Heyward, but this was also true among the white infants and children.[1] On the other hand, the slaves taken as a whole seem to have lived to a greater age than did the whites. This can be seen from the tombstones in old family burying grounds in sections of the Low Country of South Carolina, where malaria prevailed. The Negroes as they grew older became practically immune to malaria. Children suffered from it to some extent, and with infants it was often fatal. At least, in later years this has been my observation.

It is impossible to estimate the natural increase in Nathaniel Heyward's slaves, for the only records pertaining to them are the inventory and appraisement made by the executors under his will, and these show only the names of the Negroes on each plantation and their appraised value. It is very certain, however, that the increase from births was on a par with that in later years of the slaves of his son Charles, who inherited from his father four plantations and four hundred and forty-nine slaves.

[1] The authority for this statement is the late T. Grange Simons, M.D., of Charleston, S. C., for years president of the South Carolina Board of Health. He lived to be quite an old man, and in his younger days practiced on rice plantations.

These slaves came into Charles Heyward's possession during 1851, and his lists show that seven years later there were on his Combahee plantations only four hundred and forty-two, a decrease of seven. Charles Heyward had in the meantime bought a cotton plantation in Beaufort County, where it is probable that there were very few slaves and that he moved to this plantation some of his Negroes from Combahee. His lists show that, at the end of 1858, he had on all of his places five hundred and twenty-nine slaves, and his plantation records further show that, during the year, among this number there were no deaths and six births; in 1859, six deaths and six births; and in 1860, two deaths and four births, a gain of only eight during the three years. As to the births, it may be safely asserted that during the three years there were a number which were never recorded.

It is exceedingly misleading to attempt to compare the birth rate of the infants of the slaves with that of Negro infants in South Carolina today, given by our State Bureau of Vital Statistics. The Bureau records the births and deaths of every infant, but the plantation records show the births and deaths of only those who lived for a certain length of time. Why this was so can readily be explained. It was the custom on rice plantations to distribute woolen cloth during the month of November, and cloth of lighter material during May, the distribution being made by the overseer. When a baby was born, its mother, at the time of the next distribution, would take her child to the overseer and be given a yard and a half of cloth for it. She would at the same time give the baby's name, which was added to the list. However, should a baby be born and die during the period between the months of distribution, no record was made of either its birth or death, and therefore it cannot now be known what the birth or mortality rate was among the infants of the Negroes.

We do know that some of the slaves lived to be very old. Of the four hundred and ninety-five of Charles Heyward's slaves when they were emancipated, there were two hundred and thirty-three males and two hundred and sixty-two females, including children of all ages. A careful inspection of this list shows that of the males, fifty-three, or 22.7 per cent, were forty-five years old or older, and of the females, fifty-four, or 20.6 per cent. This percentage exceeds the percentage of both the white and Negro population of South Carolina, who had attained the age of forty-five years, made by the United States Census Bureau in 1925. This census shows that of the white population of all ages then living, 15 per cent were forty-five years old and older, and of the Negro population, only 14.9 per cent.

If we extend the census of the government to ages above forty-five, the ages of Charles Heyward's slaves show an even better comparison, for when his Negroes were freed there were on his Amsterdam plantation ninety-two, of whom seven were above the age of seventy; five being seventy-eight, one seventy-six, and one seventy-three, being nearly 8 per cent of the total.

There were ninety Negroes belonging to Charles Heyward's Lewisburg plantation, of whom seven were between the ages of seventy and eighty. At Rose Hill, his home, though two of the one hundred and forty Negroes living there were eighty years of age, the percentage of old ages did not run quite so high as that of Amsterdam. Still there was quite a large percentage between sixty and seventy years of age. Pleasant Hill plantation showed a percentage of old people as great as that at Amsterdam and Lewisburg.

The longevity of slaves on rice plantations can be readily accounted for. When they became too old to work, they were practically pensioned, being fed and clothed just as they had always been. They continued to live in the houses they had occupied all their lives. Had a roll been called two years after their emancipation, few of these old Negroes would have answered to their names. For many of them freedom meant starvation. Having been dependent all their lives on their owners, they could not adapt themselves to new conditions, and many of them soon died.

It is worthy of note that among all of Nathaniel Heyward's slaves, as shown in the inventory of his estate, only one, a boy on Clay Hall, one of the plantations he purchased, is noted as being a mulatto. There were a few on some of his other plantations who had Indian blood in their veins, and one of these, John Mustifer, was my head carpenter for some years.

John was a fairly good carpenter. He was smart in some ways, but tricky in others, and many of the Negroes were somewhat afraid of him. I never saw him frightened but once. Something had gone wrong on the inside of a large river trunk, and at low tide John went into the trunk, which scarcely any other Negro would have done, to see what the trouble was. It is quite dark inside of a trunk, and while he was crawling on his stomach through the trunk, he heard something also crawling through it, very close beside him. He promptly made for the opening, and as he reached it, so did a large alligator, and both plunged headlong into the river at the same time. John used to say afterwards that he didn't know which was more frightened, he or the alligator.

The highest valuation placed on any of Nathaniel Heyward's slaves by his executors was a thousand dollars, and the lowest, twenty-five. First on the list of each plantation appears the names of the drivers, carpenters, blacksmiths, and others who had trades. Next came the field hands, the house servants being listed by themselves. Negroes who on account of old age were liabilities rather than assets were valued at small amounts, as also were young children.

When I began rice planting in 1888 there were on my plantations a number of Negroes whose names I find in the inventory. Some of these Negroes had worked for my great-grandfather, and others had been too young to work for him, but this did not prevent many of the latter from declaring that they remembered him, which I knew was not true. As I now see the names of these Negroes, I well remember their faces, their individualities, their virtues, and their faults. Among them were some of the best Negroes I have ever known, some of whom, after Rose Hill and Pleasant Hill had passed into other hands, moved to Amsterdam and Lewisburg, to work for my father, whom they well knew.

Among many others in the inventory of Nathaniel Heyward's estate, I find the name of Caesar Pencile, the blacksmith at Rose Hill, in later years a great pet of my grandfather's. Caesar was then forty-six years old and was valued at a thousand dollars. And there, too, is the name of my old friend, Cudjo Myers. Whenever I see a picture of a group of Africans in their native jungles, among them I always see Cudjo.

10

THE COST OF A CROP IN THE DAYS OF SLAVERY

THE value placed upon the slaves by the executors of Nathaniel Heyward was $1,000,000. In addition he owned five thousand acres of rice lands, of which he usually planted about forty-five hundred acres, allowing certain squares to rest each year.

His holdings of highlands, on much of which was a growth of fine timber, were very extensive, but as to the total acreage of these lands it would be impossible to estimate.

Good rice lands in his day, when reclaimed, were valued at from fifty to sixty dollars an acre. The highlands brought low prices, and the timber on them was seldom taken into account—the greatest mistake our rice planters have always made. His total estate, including land, slaves, residences in Charleston, stock in banks, cattle, mules, horses, etc., on his plantation, and silver and furniture, amounted to $2,-018,000.

This was certainly a large estate for one to accumulate early in the nineteenth century, but what is most remarkable is that it was made entirely from the soil, without mules or machinery of any kind, with only the Negro, the hoe, and the rice hook. Much as I admire the ability of my ancestor both as a planter and as a business man, I must say that for many years the Lord was certainly on his side, and that He must have tempered the east winds and held back the rising tides.

I remember once asking one of Nathaniel's former slaves if his "Ole Maussuh" ever had any bad breaks in his river banks, and if so, how he managed to close them without losing his crop, and this was his explanation.

"W'en me Ole Maussuh cut rice, nigger een 'e fiel stan like blackbud, an' ef 'e hab broke fore 'e rice tie, dem nigger teck um out fore de water ketch um. We'en de broke ready fuh stop, 'e call all 'e obshur an' dem fetch all 'e nigger. Dey come spang from 'Clay Hall' an' how dem nigger meck de dut fly. Fas' as one dem trow dut in de brake, nudder trow on um. Nigger haffer wuk w'en Ole Maussuh stop broke. Dem muddy up for sough. 'E buy plantashun like 'e buy pipe."

Practically all of Nathaniel Heyward's estate was in land and slaves, and it would be interesting to know if his planting operations yielded him a fair return on his investment. It is certain that they must have done so during the first twenty-five years of his life as a planter, but in later years, when he owned so many plantations and such a large number of slaves, it is very doubtful if they did. Dr. Gabriel Manigault places his grandfather's net income in the year 1805, when his investment could not have been half as great as it was forty years later, at $120,000, giving his authority, and states that in the year 1818 his income was $90,000. Referring to these years, Dr. Manigault says, "During the European wars of that time, all of the cereals sold well, and in 1805, the year of the battle of Austerlitz, rice brought a higher price than ever before or after, up to 1860."

In 1805 Nathaniel Heyward could not have had a capital investment in plantations and slaves exceeding $1,000,000. Therefore, for him to have realized a net profit of $120,000, and a profit of $90,000 in 1818, shows that during these two years he received not only a large return from his rice crops, but in addition a handsome return on his capital invested in land and slaves.

The period from 1805 to 1815, as Dr. Manigault points out, was exceptionally favorable for rice planters on account of the high prices, and although nearly every year that Nathaniel Heyward planted rice, his profits were considerable, I am convinced that they were never as large as during the above periods, and that during the latter part of his life his profits could not have included a fair return on his investment. Probably this did not deter him from increasing his planting whenever opportunity presented itself. Few farmers today include interest on invested capital in the cost of their crops. If they did, many of them would be so discouraged that their present problem of over-production of food stuffs would be speedily solved.

As far as rice-growing sections of South Carolina and Georgia were concerned, after the tidal swamps had been reclaimed the amount of capital invested in the slaves in many cases must have rendered the system economically unsound, and this, I am sure, became more and more evident each year. In fact, if the slaves had not been emancipated as a result of the Civil War, it is almost certain that their owners would have soon begun to favor their gradual emancipation, whether or not they received remuneration for them from the Federal government. In addition, the slaveholders were beginning to realize that the sentiment of the world was turning against the owning of slaves.

Among the slaves on rice plantations, many could be depended upon to do excellent work, but taking them as a whole they could by no means have been efficient laborers. How could they have been? They received no remuneration for their labor, and whether their work was well or badly done, they knew they would be well fed, clothed, and housed. If the crop was not properly worked, that was Maussuh's worry, not theirs. This was undoubtedly the feeling among most of the slaves, although there were many notable exceptions.

There was additional reason why slavery was not as profitable as it was thought to be. In order to employ in the fields a sufficient number of laborers capable of working his crop, the planter had to support a number of Negroes who were unable, on account of old age, youth, or disability, to do any work whatsoever. This added considerably to the cost of his crop. For the planting, cultivating, and harvesting of his rice crop he had to depend wholly upon his field hands. The Negroes on the plantation who had trades, the carpenters, blacksmiths, coopers, *et al.*, never worked in the fields.

In confirmation of this it is only necessary to glance over the list of slaves on Rose Hill plantation, a list compiled at the time of their emancipation. There were one hundred and thirty-seven slaves on the plantation, and of these only fifty-six were field hands, among whom were a few children classed as half hands. Having no occupation whatever were eight old Negroes incapacitated for work, two women unable to work on account of disability, and thirty children under twelve years of age. These numbered only sixteen less than the entire number of field hands. The rest of the slaves on Rose Hill, forty-one in number, had various occupations, from poultry woman to driver. Only 42 per cent of the Negroes on the plantation worked in the rice fields.

Under such conditions, what did it cost to grow an acre of rice under the slavery system? This question cannot now be accurately answered, if it ever could have been, and for the following reason: All of the planters' business was transacted through factors, usually in the nearest city. Through these factors, they purchased nearly all of the supplies needed for their plantations. The factors sold their rice, and with them the proceeds of the sales were often deposited. The planters sometimes drew on their factors for their living as well as their plantation expenses. After the Civil War, the business of the ante-bellum factor rapidly decreased. Some never resumed business. The oldest firm of factors in Charleston failed in 1891, and, from then on, the selling of rice was done entirely through brokers.

Some of the accounts of factors with planters in the days of slavery may still be found after much research, but a close inspection of them would not enable one to ascertain what it had cost to plant an acre of rice, for, in addition to the amount expended for plantation expenses, it would have to be known how many acres had been planted. There may possibly be some old accounts in existence which show both the cost and the acreage on a certain plantation, but this I seriously doubt.

Let us consider Nathaniel Heyward's net profit of $120,-000 realized in the year 1805, and endeavor from that to form an approximate idea of what his planting expenses were for that year. He could scarcely then have planted more than three thousand acres in rice, but those acres were comparatively new lands and exceedingly fertile, more so than they were in later years, besides being naturally the best lands he owned. This being the case, a yield of fifty bushels per acre on his three thousand acres is not too high an estimate, especially when it is known that fifty years later his son, Charles Heyward, planting about twelve hundred acres, averaged for five successive years, forty-five bushels per acre.

It therefore can be taken for granted that Nathaniel Heyward harvested 150,000 bushels of rice in the year 1805. Of this amount, he probably kept on his various plantations 9,000 bushels for seed for another year, and for other purposes, and shipped to his factors in Charleston 141,000 bushels. The price of rice at that time being high, he must have sold it for at least $1.25 per bushel net on his plantations. This would have amounted to $176,000, of which Dr. Manigault says $120,000 was net profit, which would indicate that the cost of his crop was $56,000, slightly over eighteen dollars per acre. This would appear a high cost during the years of slavery, when the rice planters had not weekly pay-rolls to meet, but here a question arises. Were the living expenses of himself and his family included in the $56,000, for, if so, the cost of his crop would have been very much less than if the whole amount was for plantation expenses. Only an itemized account of his factor's books would disclose this.

I am quite sure that the cost per acre of his rice crops, and the question as to whether he was making interest on his large investment in land and slaves, never worried Nathaniel Heyward. He left no records of their cost, and I doubt if his factors knew. Through them he bought everything he needed for his plantations, and what he was most interested in at the end of the year was the balance to his credit with his factors after the sale of his rice, for if ever there was a man who kept in close touch with his business it was Nathaniel Heyward.

Never did the slightest detail regarding the work on his plantations seem to escape him. When eighty years old, writing a letter from Charleston to his son Charles, then forty-five years of age, who himself was an excellent planter and had charge, under his father, of several of his plantations, he gives most minute directions as to the hitching up of a pair of oxen. In his letter he said, "I suppose a yoke of oxen may be borrowed from the Pines [one of his plantations] and worked in the big cart, by the tongue, with leaders. This is now in operation at Marshland [another of his plantations] with an ox-cart from Kirker's shop very successfully. Observe that a small rope is hitched on the horn of the off ox and is led by the boy without lash or switch. The plowing is also done by the oxen. Observe that the Ox-cart belongs to Ashley farm [his also], though in common use for both farms at this time." Imagine one of our millionaires of today giving such directions as to the hitching up of a pair of oxen.

What were Nathaniel Heyward's views regarding slavery? He certainly must have favored the system, or he would not have owned so many slaves. But it must be remembered that in his younger days the owning of slaves was almost world-wide, and the Abolition movement, starting in the Northern states, had not begun. It must also be taken into consideration that as far as he was concerned his was the fifth generation of his name who had owned slaves in Carolina, and hence he must have regarded the institution of slavery in the light of an inheritance, one to be passed down from father to son. It can be said of him, however, that never did he sell a slave. No slave family on his plantation ever had cause to fear they would be separated. Every African brought to Combahee, everyone born there, was buried in the Negro graveyard on the plantation where they had lived.

These old slave burying grounds on the former rice plantations are still being used. In them, emancipation made no change. A Combahee Negro of today, never mind how far away he may have wandered, when he comes to die, thinks of the old plantation as his home, and wants to be carried back and buried among his own people, even though they were slaves.

Having been thrown as I was for a number of years in contact with old Negroes who had belonged to Nathaniel Heyward, I have often been struck with the fact of how highly they regarded him. They always referred to him as their "Ole Maussuh." And I have been made to feel that they derived more satisfaction from having worked for him as slaves, than they did from working for me as freedmen. Often they would brag to me about him—what he did, and all the plantations and Negroes he owned, contrasting the life he led with the life I was leading. It was often amusing to hear them giving vent to their imagination, with a superabundance of which they were endowed.

One day I shall never forget, for a former slave drew a decided contrast between the buggy in which I rode and the carriage which my great-grandfather used. I was riding through the plantation and met the old darkey on a causeway. I stopped for a little chat with him, during which reference was made to Nathaniel Heyward. "Maussuh," he said, "Wha' mek you ride in buggy? Ole Maussuh Nat, him ride in karrage w'en 'e tek 'e pleasure. De buckle on 'e hawses' harness shine lak gole, an' de karrage heself 'e shine too. W'en ona look puntop um, tink see ona self in looking glass. An' 'e hab glass en 'e window, too, dur let down. Two coachman set on de high front seat, an' dem hab on one uniform, an' all two hab high hat. One drive, de tudder fole 'e arm an' sit up straight. Nudder nigger perch 'eself on de back ob de karrage. Him too, hab uniform, an' 'e hole on wid two strap. Der karrage drive to de front do', an' ole Maussuh get een berry spry. Maussuh drive bout de whole ebenin; 'e gone spang Pines." Proceeding, he then answered his own question in regard to my riding in a buggy. "An' bless Gawd, ebery acre of lan' Maussuh's wheel turn on dat ebenin belong to Maussuh, an' ebery nigger 'e pass een de road belong to Maussuh too. 'E meet one young nigger dur walk de road. 'E stop. 'Boy, who you belongs to?' Maussuh quizzit um. 'Me belongs to you, sur,' de nigger say. 'Wha place you belongs to?' 'E quizzit um gin. 'Lewisburg.' 'Who yo Ma?' 'Judy.' 'Ona ought to be a good nigger,' Maussuh tell um, 'but you berry far frum home.' Den 'e say 'Drive on.' Dat duh *my* Maussuh!"

11

CHARLES HEYWARD'S DIARIES

NATHANIEL HEYWARD had nine children, six sons and three daughters. Of his sons, only two survived him, Charles, next to the youngest, and Arthur, the youngest. By his will he left to Arthur a number of slaves and the Bluff plantation, where he had made his home for so many years. To Charles he bequeathed four of the five best plantations he owned, Amsterdam, Lewisburg, Rose Hill, and Pleasant Hill. The rest of his estate, real and personal, he divided among his grandchildren, the children of his deceased sons and daughters.

Charles, from his early manhood, had given his attention to assisting his father with his plantations. This probably accounts for his father's giving him the four plantations he did. Amsterdam and its adjoining plantation, Rotterdam, which were called by Nathaniel Heyward his "gold mines," were the first lands on the Combahee acquired by Daniel Heyward through grants and, as I have said, remained in the possession of his direct descendants and were planted by them until the growing of rice was abandoned in South Carolina.

Charles Heyward was born in Charleston October 31, 1802. His boyhood was spent in that city during the summer months and his winters on the Combahee. He went to school in Charleston, and later attended Princeton College, but was not graduated. After the Revolutionary War few young men in South Carolina cared to study in England. In this they differed from the generation before them, when South Carolinians at one time registered at the Inns of Court in London were more numerous than students from any other American colony. Among them was Thomas Heyward, Jun., the Signer. Nathaniel Heyward insisted on his sons' visiting Europe, but none of them went to college there.

When he was quite young, Charles married Emma Barnwell, of Beaufort, where her family had been among the early settlers. She was a descendant of Dr. Henry Woodward, of whom mention has been made. Emma Heyward died at the age of twenty-nine, and Charles, like his father Nathaniel, never married again. Emma Heyward left five children, two sons and three daughters.

Upon his return home from Princeton College in 1821 Charles Heyward began keeping a diary and he continued to keep it for forty-five years. This grandfather of mine must have been a very methodical man and much given to detail. In addition, he was something of an artist, and for a number of years illustrated his diaries. On page after page, interspersed between the lines, are quaint little water-color drawings, very minute in detail and very expressive. Some of the pictures are illustrations of happenings recorded in his diary, while others evidently were suggested by events of general interest at that time. For instance, in the year 1830, he drew a picture of a locomotive and several flat-cars loaded with bales of cotton, the train crossing a trestle. This drawing was undoubtedly suggested by the first long line of railroad

ever built in the United States, then being constructed between the little town of Hamburg on the Savannah River,
opposite Augusta, Georgia, and Charleston, although he
makes no direct reference to it.

Like many others of his time, it appears from his diary that
he was greatly interested in the weather, for he refers to it
far oftener than he does to anything else. When in this diary
he mentions favorable weather for harvesting, we find a little
picture of Negroes heading sheaves of rice out of the field, or
stacking it in ricks in the barnyard. The field hands are shown
dressed just as they were, in gray-colored suits of natural
wool, and the driver in charge of them with a blue coat and
felt hat. To emphasize the prevalence of good weather he
would draw ships sailing serenely on the ocean, and when a
storm occurred, the ships are seen tossing about on the
troubled sea. As he traveled in those days between Combahee
and Charleston by carriage, the drawings accompanying the
mention of the date of such journeys (for that was what they
must have been in those days) show how the members of his
family traveled on each particular trip, his wife and children
riding in an old-style carriage drawn by two horses, he himself going ahead in a high-wheeled buggy, with a Negro
driving it, his sons on horseback behind the carriage, and a
Negro outrider bringing up the rear.

One picture is of a funeral procession, the preacher walking in front, followed by the hearse, and after it the mourners
on foot. This drawing precedes the mere statement that an
epidemic of yellow fever was proving very fatal that week
in Charleston. Any incident occurring on a hunt, whether for
deer, duck, or alligators, he was likely to illustrate when he

described the weather on the days such incidents took place, though he says nothing regarding them. One feels sure, however, in looking through his diaries, though he does not say so, that on a certain date his boat capsized, and that on another day a large wounded alligator nearly succeeded in getting into it.

Next to noting the changes of the weather, his hobby was for boats, small craft, which he built himself and used on the Combahee. Twenty-one years after Robert Fulton made his first successful venture with a small steamboat on the Hudson, he had a picture of a small boat, propelled by side wheels, which a Negro worked by means of a crank, the wheels being enclosed exactly on the order of the old side-wheel steamboats used within the memory of many today. A Negro sitting midway in the boat turned the crank-shaft by means of a lever, while a helmsman sat in the stern. A hunter, gun in hand, occupied the bow. From the picture, the boat seemed to be moving along smoothly and briskly. Two years later he must have been pretty well informed as to what Fulton had done, for he greatly improved on the Negro and the lever. The drawings then show a rather crude little steam engine installed in the boat, which we know he made himself on the plantation, for another picture represents him building it, his eldest son helping, while his wife and younger children look on.

At this period of his life, much of Charles Heyward's time must have been consumed in indulging his hobby for making and sailing boats, for in his diary one comes to a picture of the "Contrary" which carried a sail in addition to her engine, and this sail, as is also seen, one day caused her to capsize in a stiff March wind. The next ventures were the "Nonsuch" and the "Fire-Fly," the latter with a smokestack nearly as tall as the boat was long. The "Fire-Fly," however, must have been quite a success, judging by the complacent looks of two gentlemen in high hats, who are shown comfortably seated in the boat, and the size of the wave in its wake. From the amount of black smoke pouring out of her smokestack, one could well imagine, too, that small as the boat and her engine were the black engineer had crammed the furnace full with the fattest kind of "light-wood" pine.

These small boats, propelled by steam, were used only on the Combahee during the winter months for duck and alligator shooting, and do not appear to have ever been taken to Charleston. The diary, however, shows that in the summer Charles Heyward did some boating, for in the year 1826 is found a pretty picture of the schooner, "Margaret." This picture comes immediately after the following entry, which is interesting if for no other reason than that it shows the cost of entertaining one's friends over a hundred years ago:

Expense of a frolic on the "Margaret" on the 5th. of August '26, the number of gentlemen present, 18.

Four bottles of claret	$4.00
1 doz. bottles porter	3.50
One hundred cigars	2.00
Dinner found by Charles including ice, etc.	20.25
17 bottle of wine and 3 of brandy	35.00
	$64.75

Three bottles of wine, one of claret and three of porter were not used.

Despite the amount of liquor on the "Margaret," however, the picture, no doubt drawn the night of the cruise, is exceedingly creditable.[1]

During the forty-odd years that Charles Heyward kept his diary, no decided change in the weather took place which he did not mention. No late or early frost occurred, no unusual cold or excessive heat, no undue amount of rainfall, nor was there an extended drought, which is not commented upon. This interest in the weather must have been general in this time, for it is known that many kept similar records. The newspapers of Charleston then carried, as they do now, notices of the weather, but these notices told what the weather conditions had been the day before, and were not, as they are today, predictions of what might be expected tomorrow. Correspondents frequently wrote letters to the papers in regard to the weather, and, judging from the number printed, the newspapers were glad to publish them. Such letters, no doubt, Charles Heyward read with much interest, and a few of them he preserved by putting them in his diary.

Among them was one written during 1832, to the *Charleston Courier*, and signed "D." It began thus: "Mr. Editor: We have had indeed a pleasant month of June, and everybody says we have never before had such cool weather in that month. I will show the month of June for 6 years." "D" then proceeds to give the daily ranges of the thermometer for the month of June of the six preceding years, a most

[1] Pages from the Diary form the end papers of this volume.

formidable array of figures. As to his temperatures for one of the Junes, however, Charles Heyward evidently differed with him, for he attached a letter to a page of his diary and pasted a little notation on it, stating: "June 1828—Don't correspond with my Diary."

In little leather-covered books, one for every plantation, he kept a record of each year's crop, the number of acres planted in rice, the dates of turning the land, planting, hoeing and harvesting, the yield per acre, and, for some years, the price per bushel which his rice netted on the plantation. But to me the most interesting record he kept is the one concerning his slaves, the one which, until well into my own time, still was called "de book" by his former slaves, few if any of whom are now living.

"Do, Boss, get de book and gimme me age," has been asked of me many times on the Combahee. "De book" is an ordinary blank book, about eight inches wide and fourteen long, with blue pages, and its heavy paper cover is a much darker shade of the same color. The first entry in this particular book was made December 8, 1858, and is a statement of the amount of cloth given on that date to each slave.

The first name entered in the list is that of Ishmael, head driver on Amsterdam, and below his are the names of all the other Negroes on that plantation arranged in accordance with their ages, the names of infants being at the last of the list. There is a similar list of the slaves on each of the other plantations, and opposite each name is set down the number of yards of cloths distributed semiannually.

After his Negroes were freed, Charles Heyward proceeded to compile a very exact list of his emancipated slaves, dating it July, 1865. And for this he used the same blank book he had used for other lists. Probably at that time he could obtain no other. He arranged it as he had previously done, by plantations, listing each in the same order. He then gave the age of every Negro at that time, and also the kind of work that he or she had done. If certain Negroes had been too old, too infirm, or too young to work, they were put down as having "no occupation." To show how closely he kept in touch with his slaves, not only did he give the age of all without a single exception, but by a certain mark indicated which Negroes had been left on Combahee when the others were taken to the interior of the state when the Civil War began, and by another mark indicated the few who, during the four years of war, had escaped.

To have been able to give the correct ages of all his slaves, some of whom were much older than he, proves conclusively that a complete record of all slaves must have been kept, not only by him, but also by his father before him. If Nathaniel Heyward did not know all of his two thousand slaves, his son Charles knew all of his five hundred, for he certainly wrote their names often enough.

Some very peculiar names appear on these lists of Negroes. The Negro mothers often asked their master or members of his family to name their children for them, but many named them themselves. The names usually indicate who did the naming. "Binky," "Lindy," "Sukey," and "Friday" we know were almost certainly named by their parents, whereas, those bearing classical and biblical names, as many did, were in nearly every instance named by their owner or overseer, who must often have been hard put to it to find new names.

No surname appears on the lists, although a few of the most prominent Negroes on the plantations, such as old blacksmith Caesar, were known by both given and surnames. It is well known, however, that among themselves the slaves all had surnames, and immediately after they were freed these names came to light. The surnames were selected by the Negroes themselves. Scarcely ever did a Negro choose the name of his or her owner, but often took that of some other slaveholding family, of which he knew.

For years after the Civil War, the foremen who had been drivers, and the other old Negroes, liked to pretend, especially in my presence, that they did not know the surnames, or the "titles" as they termed them, of the younger and more trifling of the hands they worked. By the tone of his voice, an old foreman would plainly insinuate that a Negro boy or girl had no right to a surname, at least to one that he was required to recognize in order to keep the time-book. He had not been known by his when a boy. "Boy, wha yo' title?" I have often heard a foreman ask, in a seemingly contemptuous and angry tone, of a small, very black half hand, who, grinning from ear to ear and showing his large white teeth, would answer, "My title is Monday White." I was certain all the time that the foreman knew what it was, but did not want to admit before me that he did.

12

ROSE HILL PLANTATION

FROM early manhood, Charles Heyward considered Rose Hill plantation his home and lived there until the Civil War. Rose Hill adjoined the Bluff, and the residences of Charles and his father were only a short distance from each other. A canal, along which grew a line of cedar trees, separated the two plantations. The canal and the cedars are there yet, but of the two houses, nothing remains. At Rose Hill, only a walkway of flagstones, which led from the front steps of the residence, indicates the exact spot where it stood.

On August 1, 1835, Charles Heyward in his diary drew a picture of his home in Charleston, located in one of the uppermost wards, then known as Wraggborough. In fact, it was so far from the center of the city that he refers to it as being located on "Charleston Neck." The picture shows a two-story white house built on a high, arched, brick foundation, with a piazza running the entire length of the first story and a high flight of steps leading up to the front door. On the second story, above the front door, is a triple-arched window. The roof of the house and of the front piazza are painted, as are also the shutters; and the lot in which the house stands appears to be quite an extensive one. A picket fence separates the front and back yards.

On the death of his father in 1851, Charles Heyward fell heir to the home in Charleston, but he makes no reference to this, simply heading the pages of his diary, "241 East Bay," and then he comments on the coldness of the weather. In this house he lived approximately ten years.

Although Charles Heyward was nearly fifty years old before his father died, he had by no means lost his love for boats, for on October 23, 1850, he cut the following notice from the *Charleston Courier,* and pasted it in his diary, "Launch—A fine new schooner, built by Messers Addison & McIntosh for Nathaniel Heyward, Esq., intended for the coasting trade, and commanded by Capt. J. B. Morgan, will be launched, full rigged, from their Ship Yard this morning at 9 o'clock."

This schooner was named the "Acorn" and later became the property of Charles Heyward, carrying his rice to Charleston for eleven years. His diary shows that he sold the "Acorn" on March 15, 1862, for he could use her no longer after the Civil War began.

Although I cannot state it positively, I am sure that Captain Morgan of the "Acorn" was the first and only white captain in charge of any of the schooners owned by either Nathaniel or Charles Heyward. All of the other captains of their schooners were their own slaves. Morgan, however, did not remain captain on the "Acorn" very long, for Charles Heyward's records show that one of his slaves, Richard, succeeded him, and remained in charge of the vessel until it was sold.

Viewed from the adjacent highland, the rice field at Rose Hill plantation, lying in a bend of the Combahee River, stretched away almost as far as the eye could see. The river bank, protecting the field, could be dimly traced by a few trees and a growth of cane. In the marsh just outside the bank there stood, at intervals, great cypress trees, their long limbs drooping gracefully, where many birds congregated during the day and some roosted at night.

Perched on the very top of the tallest of these trees, bald eagles could often be seen, calmly surveying the surrounding country, while they rested their wings before returning to soar among the clouds. For many years the oldest trees had grown on the river's edge, and generations of eagles had rested on their tops. From the size of some of these cypresses, one could well imagine that if the ships of Velasquez de Ayllon had, more than three centuries before, pressed farther up the Combahee, or the Jordan, as Velasquez named the river, when for the first time a white man set foot upon the soil of Carolina, the eagles of that day, from their eyries in these trees, must have looked down with wonderment upon the quaint ships of the Spaniards, while the cool shade of the trees protected the bold mariners from the summer sun, as they fraternized with the Indians.

At Rose Hill plantation, on the highest part of the highland, quite near the river and overlooking the rice fields, there stood before the Civil War an old dwelling house, painted white with green blinds. Not a great pretentious mansion designed by an architect following a style of long ago, but just a plain simple country home, like many others of its kind to be found on rice plantations in the Low Country.

In front of the dwelling on Rose Hill plantation a lawn,

green with rye in winter, sloped to the river's edge, and there a few live oaks, their leaves green throughout the year, spread their wide extending limbs, from which hung long streams of gray moss swaying gently in the wind. On the left, at the edge of the lawn, bordering the river, lay the barnyard, where stood the threshing mill with its tall square chimney. The schooner "Acorn," succeeding the old "Elizabeth," which had carried many a crop from Rose Hill to Charleston, could often be seen moored to the wharf, while the crew, consisting of the captain and four sailors, all of them slaves, were busy lifting from her hatches large boxes of woolen cloth and shoes imported from England through Charleston. Soon she would be loading the first cargo of rice to be threshed from the great ricks, which stood side by side in the barnyard with their narrow, peaked tops and broadening sides, thatched with sheaves like a roof.

To the left of the house was the coach house with a gabled roof, the stables, and the barn. Several servant houses were back of the "big house," facing the same way, and within and before their doors Negro children could be seen at play. In the more distant background, across the road and cornfields, were the two settlements. These consisted of rather small whitewashed houses in two rows, facing each other, with the street between them thoroughly swept and as clean as a city boulevard. At the head of each street was the house of the driver, a little larger and somewhat better than the rest. Not very far from each settlement was the nursery. Lizette, one of the most intelligent and trustworthy Negroes on the plantation, was in charge of one child house, and for many years the other, at Pleasant Hill, had been under the supervision of old Mary, equally reliable. For these nurseries, two special cooks were provided.

Some little distance back of one of the nurseries was the hospital, or sick house, as the Negroes called it, a building of good size with several rooms, neat and clean and containing beds and bedding. Here Clarissa, the plantation nurse, could always be found. And here also were the headquarters of Hannah, the plantation midwife. Beyond the settlement was another cornfield, which reached back to the blue pines, where began the great primeval forest.

Throughout the forest many cattle grazed the entire year upon native grasses and the short cane which kept green during the winter months. Droves of hogs also ranged in the woods. Stepney, the hog-minder, who years later worked for me, followed them closely and several times each year penned and counted them. Both cattle and hogs were far outnumbered by the red deer, for these lands were once the favorite hunting grounds of the Indians, and on them many deer were killed and their hides sold to traders and exported. But many deer still remained; their tracks could be seen everywhere, and also the paths they made leading to the fields where at night during autumn they often fed on the peas growing in the corn.

A broad piazza extended the entire length of the Rose Hill house, and from this piazza could be seen plainly the rice fields and the barnyard and a part of the highlands. From its western end, both of the settlements were in sight; and thus the owner of Rose Hill could always know what was taking place on the plantation. He could not only see the hands at work in every part of his rice fields, but in the squares nearest the house could note the progress they were making. The driver, as he walked through the fields inspecting the tasks, might readily be distinguished from the others by his blue suit and his hat, for all the field hands in those days dressed in light-colored clothes, and the men wore caps. When the Negroes worked in these near-by squares, they always thought "Maussuh" was watching them. "Maussuh, him hab 'e eye puntop ona [you] all de time," they would say, but what they pretended to fear most of all was a certain pair of spy-glasses. "W'en Maussuh trow 'e eye tru dem ting 'e fetch 'eself nigh for we. Dem glasses 'e pint out ebry spec' of grass de hoe lef'. De grass 'e git by March [the driver] but 'e ain't git by Maussuh."

Familiar as I am with all Charles Heyward's plantations and having known a number of his slaves and the nature of work they performed, I believe that I can reconstruct with fair accuracy the scenes on Rose Hill plantation on certain days referred to by my grandfather in his diary, even if they occurred so many years ago.

Let me select at random May 20, 1858. The plantation itself must have looked then very much as it did when I knew it, except that the dwelling and the threshing mill had not been burned, and there were a few more houses in the settlements. The child house and the sick house, I must confess, would have looked strange to me, for they were not there in my day. They belonged to the days of slavery.

According to the diary, the field hands were hoeing rice in the large Mill Pond Square. The day itself was bright and clear, with a westerly wind blowing. It was fine weather for hoeing rice, for that afternoon at 3 P.M. the thermometer had climbed to 83 degrees. The spring must have been rainy, for the hoeing of the first planted rice was about ten days late. The full force of hands from both Rose Hill and Pleasant Hill were at work, for the rains having caused the grass to grow, the sooner it could be chopped out, the better it would be for the rice. March, the head driver at Rose Hill, and John, the driver for Pleasant Hill, were walking through the tasks, seeing whether any grass was being skipped and noting how deeply the ground was stirred. I never knew these two drivers, but I have often wished I had them when my hands were hoeing rice.

Glancing over the list of Negroes who must have been hoeing rice in the Mill Pond Square that May day, I find many whose names I have often written in my time-book, and their names recall to me some well remembered faces. And though these Negroes would have then looked to me much younger, I believe I should have known them.

Among them were Abby Manigault and Frankie Shilling, the two best women field hands I have ever known, and both of them as half hands had worked for Nathaniel Heyward. When I planted rice, Abby, for years was my "foreman" of the children. No one could equal her in managing them, especially when it came to making them load a schooner. They used to carry the rice in baskets from the rice house and dump it in the schooner's hatches, and Abby never let them linger either in the rice house or over the hatches. I have a picture of her standing on the wharf in the midst of twenty children with their buckets on their heads.

The schooner, "Sallie Bissell," of Charleston, named for the wife of a large rice planter on the Combahee, carried the crops from my plantation for twenty years. She held thirty-five hundred bushels of rough rice, and many a little darkey dumped his or her basket into her hatches. The distance they carried the rice, from the rice house to the schooner, was not more than thirty feet, and old Abby kept them moving. She always carried a bunch of switches in her hand, with which she constantly threatened them, but I have never seen her use one.

Before the "Sallie Bissell" was loaded, some of the children would get pretty tired, but not Abby. Late one afternoon when the schooner was being loaded, I was standing on her deck when a girl, dumping her basket into the hatches, looked down them to see how near the rice was coming to the top, and I heard her exclaim, "My Gawd! De Sallie Brittle hab a mighty big belly!"

Imprudently one year I planted several acres near the Negro settlement in watermelons. As soon as they began to ripen I noticed they also began to disappear. I noticed, too, that the stomachs of the Negro children were becoming daily more and more distended. I told Frankie that I was going to poison the melons, and that she had better warn the children.

"Don't do dat, Maussuh," she said. "Better go git ona gun right now en shoot dem little nigger, for ona know dem gwine eat dem watermillions." And they did.

I am quite sure that if Abby and Frankie, and some of the other slaves whose names I recognize on my grandfather's lists, did their work for him that day at Rose Hill as well as they did it for me twenty-odd years later, the drivers found no grass skipped between the rows in their tasks.

From a painting by Theus

DANIEL HEYWARD

From a miniature by Charles Fraser

NATHANIEL HEYWARD

From a miniature by Fraser

CHARLES HEYWARD

From an old photograph

EDWARD BARNWELL HEYWARD

SOME TOOLS OF THE DAYS WHEN RICE WAS PLANTED IN THE
LOW COUNTRY. THE HOE AT THE UPPER LEFT WAS MADE IN
ENGLAND. AT THE BOTTOM IS A BRANDING IRON.

A SLUICE-GATE OR TRUNK USED IN THE
IRRIGATION AND DRAINAGE OF RICE FIELDS

A STUDY IN LOCAL COLOR

From an old photograph

A GROUP OF GULLAH NEGROES

A NEGRO COOK IN THE KITCHEN OF AN OLD PLANTATION HOME

ALWAYS READY FOR A GAME

BORN TO BE FISHERMEN

From an old painting

ROSE HILL PLANTATION, HOME OF CHARLES HEYWARD FOR MANY YEARS. THE HOUSE WAS BURNED BY STRAGGLERS FROM SHERMAN'S ARMY

LONGVIEW, FORMERLY HAMBURG, BUILT SOON AFTER THE CIVIL WAR

From an old photograph

BRINGING DIRT FROM VINEYARD RESERVE RICE FIELD
TO MEND A BREAK IN OLD GROUND SQUARE, ON THE
SWAMP PLANTATION

13

THE RETURN OF "OLE MAUSSUH"

IT WAS midday on November 10, 1859, and since early in the day the Negroes in the settlement on Rose Hill and Pleasant Hill had been saying, "Maussuh gwine come to-day." They had looked forward to his arrival since the head driver, March, standing in the door of his house at the head of the street in the Rose Hill settlement, had blown his horn an hour before daylight, and it was still the topic of their conversation when in the fields they ate their noonday meal.

The tenth of November had long been a day of moment at Rose Hill, for on that day "Maussuh" usually returned from his summer residence in Charleston. Scarcely ever had he come on the ninth day of the month or the eleventh. Leaving the plantation on the tenth of May, he could with certainty be depended upon to return six months later to the very day, and almost to the hour. Rain or shine, it mattered not, his carriage would drive to the door.

Until that year, Charles Heyward had driven from Charleston in his carriage, but a railroad had just been completed between that city and Savannah, and he preferred it, slowly as its trains ran, to the drive of sixty miles and the ferrying of several rivers. And taking the train to Combahee for the first time that morning did not seem to have made as great an impression on him as did the fact that having sent his carriage on a day ahead he was compelled, as he recorded in his diary, "to leave my door in a public carriage the first time in my life."

Yet, were one to judge by the contract which he had recently made with the president of the new railroad, in granting it, free of cost, a right of way for four miles through his pastures, he must have expected to be pleased with the change, for in the agreement it was expressly stipulated that "upon request, at the crossing of his avenue, any and all trains should stop for him and for the members of his immediate family, for all times hereafter." But in this Charles Heyward thought not of his personal comfort and convenience only, for he required and put in the bond that on his land "no trains should loiter" and "no wells be dug." It must be presumed from these latter stipulations that he did not wish his cattle frightened for an unnecessary length of time, nor his hogs to meet untimely ends. It may be further noted that in the agreement he did not seek to nail, as it were, these rights to the land, as is usually done in such cases by lawyers using the words "heirs and assigns." All he sought was to secure these rights for himself and the members of his family for all time. Had his people not always owned the lands, and would they not continue to own them "for all time hereafter"?

Despite the drawing in the diary of the family carriage, slowly making its way from Combahee to Charleston, it had really not taken Charles Heyward very long to make the journey, for Jack, his coachman, was a fast driver, I have always heard, and used to boast that he could put his "bittle hot een 'e bucket een town and eat dem on Combee fore 'e cole." Leaving his door that November morning in a public carriage and taking the train for Combahee, I am sure did not greatly shorten Charles Heyward's trip, for on the newly constructed railroad it required nearly four hours for the train, drawn by a quaint little locomotive burning wood and with a smokestack nearly as large as the engine itself, to reach Combahee, and nearly twice as long to land its passengers in Savannah.

A train was daily run each way, passengers being ferried across the Ashley River at Charleston and also across the Savannah when that river was reached.

The Charleston and Savannah Railroad, as it was first called, followed the seacoast, which lay about twenty-five miles to the southward, and nearly all of its length passed through a pine forest belt. Often, however, the forest was intersected by large swamps which bordered rather deep rivers, all bearing Indian names, and many of the newly constructed stations were named for these rivers. Travelers in those early days of railroading in South Carolina breathed easier when their train, passing through the swamps and across the rivers, had gotten safely over the frail looking trestles and high wooden drawbridges. About every five miles the train stopped at a little station which consisted of a high platform, a small station house, and, at some, a wood rack for replenishing the engines. These high platforms, built for the first time, I believe, by the Charleston and Savannah Railroad, continued in use until my day. Passengers getting on or off the trains did not use the steps of the car, as they do now, but walked a plank reaching from the car platform to that of the station, these being on a level.

Often, for long distances, one as long as seventeen miles, the track ran so straight that at night the headlight of the engine could be seen a long time before the arrival of the train. From the car windows during the day, the passengers often saw flocks of wild turkeys feeding in the open woods, and deer, single or in herds, crossing the track in front of the train. Where the road ran through a pasture, cattle congregated on the track during summer nights, and the engineer would have to stop his train and send his fireman to drive them off.

From the train, in those days, the country looked almost uninhabited, but this was offset in the spring of the year by the natural beauty of the woods, and especially by the various shades of green in the river swamps. There was so little sign of human habitation and of agriculture in the territory through which the train passed that it was hard for one to realize that only a few miles distant, lower down the rivers, lay many of the largest and most prosperous rice plantations in the world. At the same time many thousands of acres of highland as fertile as could be found anywhere, much of it original grants to the owners of the plantations of which these lands were a part, not only lay uncultivated but were primeval forests. Being unsuited for the cultivation of rice, they were considered of practically no value except for pastures and hunting. The lack of drainage, which could readily have been obtained, together with the close proximity of the Negroes on the plantations, first made this fine belt of highland, stretching throughout lower South Carolina from the North Carolina to the Georgia line, undesirable to small white farmers, who did not wish to come into close contact with the Negroes. Slavery thus produced one of its most baneful and lasting effects upon much of the richest territory in the Southern states.

As the hour for the arrival of Charles Heyward approached, the most prominent Negroes on Rose Hill and Pleasant Hill, and also his house servants, would gather in a group near the front steps to "tell um howdy." All were neatly dressed. Most of the men wore felt hats, while the heads of the women were covered with colored plaid handkerchiefs. Two of the men had very white heads, for old Buck and Tony, both then past seventy, quitting their work in the garden for a while, had come with the rest. All of the women wore white aprons; the two seamstresses, Katrina, the younger, somewhat stout (or rather she was quite so later in life when I knew her), and old Doll, wearing "specs," looked perhaps a little neater than the rest. Among the others was Polly, the family cook, in the prime of her life, and looking the part. Tom, the butler, alone remained in the house. Slowly, and with a certain amount of dignity in his tread, he walked from one end of the lower hall to the other, entering each room as he passed, scrutinizing for the hundredth time every piece of furniture and every old gilded picture frame. Discovering a little dust under or in the carving of a piece of furniture, he promptly called the reluctant Flora, the house girl from the front, "Look yeh, gal, wha you mean; who tell you fur stop? You so everlastin' triflin', so lub fur set down all de time. Wha Maussuh gwine say? Cum yeh; go fetch yo rag." Often he looked through the back door toward the road to see if the carriage was coming, for he must be the first to greet his master and welcome the family home, as he had done for years.

A little apart from the others and near the edge of the circle stood blacksmith Caesar, Caesar Pencile. He held firmly in his right hand before his body a somewhat worn beaver hat, given him by his master, and he carried a peeled hickory stick with many notches. His coat was long and rather faded, and in his left hand was a bright handkerchief. Caesar was a Negro of medium size and slightly built, very black, but with straight and delicate features, and with hands so small one wondered how he could perform his work. In his manner he was as gentle as a woman, and no Frenchman ever surpassed him when he bowed. Dignity and also respect showed in his every gesture, and when he spoke one felt that he told the truth. Respected by master and overseer alike, Caesar required for himself respect from the other Negroes and often sought to teach the young ones "manners." His two sons, Albert and Ephraim, to whom he taught his calling, were almost as polite as their father and had many of his ways.

Blacksmith Caesar lived to be over eighty. I knew him well, and can well remember the old hat which he had for thirty years. The silk had nearly all worn away, but the frame still stood upright, and the old man carried it as he had in days gone by. On Sundays I would sometimes meet him going to church, and he would stop, holding his hat in his hand, and inquire after the health of "de young Boss en de Missis en de leetle Missis." Poor old Caesar, he outlived his day and generation. He saw Rose Hill pass into other hands, and yet there the old man stayed until he died, depending upon the charity of others who had little to give.

Further away, in a group by themselves, were the carpenters, seven of them, all young men, with Boston, head carpenter, a little to the forefront. And among them was another Caesar, Caesar Goff, as he was afterwards known, then only thirty-eight years old, a small Negro with a good, open countenance but not overburdened with sense, though faithful always and entirely honest. In fact, he was so honest that frequently in later years, when working for wages, he would "condemn" his work when improperly done, as was often the case, expecting to be paid for it however, as usually he was. Anyone knowing Caesar later could well understand why Boston, four years younger, was put over him, for never could Caesar have given directions, and both planter and overseer in those days knew the capability of each slave. But of all the Negroes there that day, none was gladder, I know, to see his master return, than Caesar Goff. His heart never changed; he was loyal to the end and never seemed to realize that he was free.

Conspicuous by his absence, was March, the head driver, who was responsible to the overseer for one hundred and fifty slaves. March was then in his prime, being in his forty-third year, tall and straight and of powerful build. Intelligence and understanding showed in his face, and force in his every movement. He looked indeed like one born to command, and it was said of him that no better driver ever stepped across a quarter-ditch. That day March and his gang were breaking corn, and in the distance he could be seen in their midst, watching them closely and giving an order every now and then in a tone that compelled obedience. He knew "Maussuh" would look that way as the carriage passed, and he knew further that it would greatly please him to see the gang at work and the corn, slip-shucked, piling up between the rows.

At last the carriage could be seen driving across the causeway of the back fields, from the little weather house just built beside the railroad track, with Jack sitting very straight on the box; and as it rose to the higher ground, a most pleasing sight must have greeted the eyes of the planter. He saw his Negroes at work harvesting his crops; he drove through his plantation looking just as he had left it six months before; saw his old home, and before it the green lawn and the live oaks. Beyond lay the barnyard with great ricks of rice, safe from storm and freshet, waiting to be threshed. Spread out before his view, the rice fields stretched away to the river's bank, water showing here and there.

As he approached his front gate, it was opened wide by several little Negro boys, who had been waiting for hours. Jack swung around the circle and with a word quickly stopped before the house. The carriage door was opened by Tom, the butler.

As Charles Heyward alighted, he saw his slaves waiting to greet him, and the first to whom he spoke, I am sure, was Caesar Pencile. I know he took the blacksmith by the hand and all that Caesar could say, with a beaming smile and with deep feeling in his voice, was "Maussuh, my Maussuh." One after another the planter spoke to the other men, who had taken off their hats as the carriage approached, and had come forward to meet him. He called each of them by name, and of each made some personal inquiry. The women, though they appeared more anxious to speak, held back and still stood in a little group, but when he came to shake hands with them, they showed their feelings more than the men, curtsying low and exclaiming, "Huddy, Maussuh, Gawd, Maussuh look so fine. How ona ben, Ole Maussuh?" "Me so glad fur see you come." "How de family?" "T'ank you, Maussuh, I berry poo'ly." This last word from old Doll. All of the Negroes were much pleased and a little excited, too, at seeing him again, and for a while some of the men lingered on the lawn.

At Rose Hill plantation that evening, while the profound stillness of the night was broken only by the noise of the ducks in the fields and the barking of a few dogs belonging to the Negroes in the settlement, Charles Heyward sat in his library at an old rosewood desk, which had been used for fifty years by his father and is still in the family, and went over with the overseer everything that had happened on the plantations during his absence: the harvesting of the crop, the condition of the place, and the work and conduct of the slaves. Carefully he jotted down the births and deaths which had occurred during that time. When the overseer had gone, from a cubby-hole in the desk he took a small yellow leather-covered book, like those in which he had kept his plantation diaries for nearly forty years, and recorded the events of the day, among others the fact that for the first time he had returned home by rail, and had that morning driven from his house in town in a "public carriage." Later, alone in his house, the planter slept peacefully, his doors and windows unfastened, fearing no harm from his hundreds of slaves, but rather putting his trust in their faithfulness, knowing that their presence was a protection and not a menace.

In less than six years after Charles Heyward came to Rose Hill in November, 1859, great changes had taken place, as a result of the Civil War. In fact, as a plantation, Rose Hill had ceased to exist. Several breaks had occurred in its neglected river banks, and the tides flooded the rice fields at will. No crops were grown. On the highlands, the fences were down, and all of the cattle and hogs had mysteriously disappeared. Fire, as well as water and depredation, had done its work. With the single exception of the houses in the settlements, nearly all of which were empty, all buildings of every description were burned—barns and stables, blacksmith and cooper shops, even the hospital. In the unused barnyard tall weeds grew, and of the threshing mill only the square brick chimney remained.

14

HUNTING ALONG THE COMBAHEE

IN THE years now long past, there was no better or more varied hunting to be had anywhere than on the rice plantations along the Combahee, and even today, where these lands have been protected, some game still remains.

Charles Heyward, like most rice planters of his day, was fond of hunting, and if old Adam Morgan, his "deer driver," used to tell me the truth, he much preferred deer hunting to any other kind. He always kept, according to Adam, a fine pack of hounds and killed many a deer. I can very distinctly recall his old Wesley Richards muzzle loader, its stock with rows of small ivory pegs, each representing a deer he had killed. What became of the old gun, I do not know. Adam used to claim that he had driven through his "stand" every deer marked on the gun, and this was probably true.

I used to have quite a fancy for old Adam, who was a most enthusiastic talker, especially on the subject of deer hunting, though how great vent he gave to his imagination it was sometimes hard to determine. He was a very black Negro, of medium height, rather slightly built, strong and wiry. To the day of his death he fully enjoyed the reputation he had made as a deer driver, and the old Negroes on the plantation used to say of him that, when he was stopping the hounds, he could, hat in hand, gallop his "marsh tackey"[1] through thickets so dense that a rabbit could scarcely get through them. Nothing, they said, ever stopped him; fallen logs, briar patches, cane breaks—without slacking his speed he would go through them all. "W'enebber him haffuh stop dog him jus' ent gib a damn. Not dat ole nigger."

Adam lived to be a very old man, but he never forgot his calling and delighted in talking of his hunting days. He pretended to remember not only the name, but the color, traits, and tongue of every hound that had followed him nearly half a century before. I am quite sure the old darkey in his dreams often hunted deer, as, one imagines from his muffled barks, a dog dreams he is hunting as he sleeps before an open fire on a cold winter night.

For all kinds of game the hunters in the Low Country of South Carolina used twelve-gauge, muzzle-loading shotguns, of the best English makes, such as Purdy, Wesley Richards, Greener, and Lancaster. In keeping with their guns were the metal powder-flasks, and leather shot-pouches with accurate measures, both of which, suspended by a cord over their shoulders, hung by their sides. A little metal capping device was sometimes used to expedite putting on the brass percussion caps, and the wads for their guns were carried in the right-hand pocket of their hunting coats.

Hunting deer was the favorite sport of the rice planters of Charles Heyward's time. There were so many deer, they were so close to the homes of the planters, the "drives" were

---

[1] A small horse said to have been brought by the Spaniards to the Low Country of South Carolina and Georgia, and bred to stock imported from England.

so short, and the hounds so eager to hunt—all of these must have been a great temptation. Next to hunting deer they preferred shooting partridges, large coveys of which were to be found in the pine woods which were unusually open, affording excellent shooting. On the highland, skirting the rice fields, coveys of partridges could always be found. Seldom did these range far out in the rice field, and when flushed they always made for the woods close by, where there were many thickets, and it took a hunter who was especially good at snap shooting to make a day's bag.

There was another reason, I think, why the rice planters on the Combahee preferred hunting deer to partridge shooting, which was that they could keep their hounds on their plantations during the summer months much better than they could keep pointers and setters. Hounds, if occasionally exercised by the driver, could be penned up the entire summer, while pointers and setters under such circumstances would often die. The bird dogs they did have were usually well bred; some of them I have heard were imported from England, the summers of which they could stand better than those of the Low Country of South Carolina.

During the months of January and February, numbers of snipe frequented such back fields as were not well drained during these months. These fields were known as "snipe bogs," and in them, on a cold day, one could shoot snipe for hours, for when they were started up, they would circle around the field, fly high for a while, and returning light a short distance from where they had been flushed. In the bogs they could be walked up in rapid succession, first zigzagging as they uttered a little cry, then flying straight away, presenting a most inviting mark if the hunter waited until they did so.

There were woodcock also in the rice country, far more in those days than there are now. They came in flights, and some winters their flights were very heavy. Unlike snipe, these birds were far more difficult to kill, for they were to be found only in the woods, in low, wet places where the undergrowth was heavy. They would rise noiselessly, springing, it seemed, from the ground, and fly straight upwards with a sidewise motion and flapping of their wings. For these reasons, hunters rarely made full bags of woodcock, but those they did kill were fully worth the effort.

There was a rule in deer hunting in the Low Country of South Carolina which was always insisted upon—that when a hunter was on a stand he should shoot at nothing but a deer. There was one exception, however, to this rule. Should a flock of wild turkeys pass within gunshot, he would be excusable if he shot one, even if the hounds were heard running a deer straight to his stand.

On the Combahee there were once many wild turkeys, and there may still be a few there. These magnificent birds are natives of America. Perhaps they antedated the Indians. Certainly for centuries they antedated our first white settlers, who exported many of them to foreign countries. I shall never forget my first experience with a flock of turkeys shortly after I began planting rice.

One of my rice fields, containing a hundred and forty acres, was separated from the rest of the plantation by Cuckols Creek, across which Negroes and mules had to be ferried in flats. No one lived within a mile or two of this

field, and seldom did anyone pass through it except when
work was being done. Back of the field lay an impenetrable
jungle, generally known as a bay, through which it was practically impossible for one to make his way.

Often in the fall when the rice had been harvested, while
walking through this field, in the early morning or late afternoon, I could see a large flock of turkeys feeding in the field
but never advancing into it more than a hundred or two
hundred yards from the bay, and as soon as they saw me
approaching, they immediately returned into the thicket. I
determined I would kill one of them, and one morning,
getting up before daylight, I hid in the edge of the bay and
awaited developments.

Just as the rising sun began to light up the eastern sky, a
large gobbler, the most magnificent bird I have ever seen,
came strutting out of the bay into the field, just beyond gunshot from where I was concealed, and strutted around all by
himself in the open. The rising sun glistened on his burnished
plumage and caused the bars on his wings to shine like golden
epaulets.

After doing this for a short while and assuring himself
that everything was safe, he began to gobble, a signal for
the rest of the flock to come out to him. From the thicket
another gobbler answered his call, and then another; soon
several gobblers joined him and all gobbled together. They
were followed by a flock of about thirty hens, running into
the field to join their lords. All of them then fed together,
picking up the shattered grains of rice, but the gobblers
seemed alert, and ready at the first sign of danger to return
to the thicket.

For at least an hour, I lay in my hiding place watching them, and then I heard voices in the distance; the plowmen were coming to work, and I saw the turkeys start back to the bay. I tried to cut them off, but as soon as they saw me, they rose in flight and disappeared into the woods. To this day, I am glad I didn't kill that gobbler. He was too fine a bird, too great an old aristocrat.

But that fine flock of turkeys was doomed. The great storm of 1893 flooded the bay to the depth of several feet. The velocity of the wind prevented them from taking refuge on the limbs of trees, and in the darkness of night they were drowned. Never were they seen in the field again.

What today would be considered the finest sport was the duck shooting on the Combahee River as it was in former years, for then the rice fields of South Carolina and Georgia, during the late fall and winter months, were the favorite stopping places and feeding grounds of these birds, which, following the natural instinct of their kind, annually winged their way from the lakes and marshes of Canada to the sunny water of Louisiana and on farther south.

On a November day, shortly after the season had begun, when an east wind, cool and damp, told of the coming of winter, often have I sat in a boat behind a blind, watching against the gray sky long lines of ducks headed southwestward. As they passed over the field, often a flock would seem to halt and hesitate and then circle slowly, spiraling downward like aerial pilots, searching for a landing place. Finding it at last, they paused. An old drake with steel-gray body and with bright green head outstretched, poising on drooping wings, would hesitate no longer. Pointing his wings downward, he would drop to the water, splashing as he struck. Quickly the others followed and, hidden in brown stubble, thrust their heads beneath the water and fed on the grains of rice.

All day the whistling of fast-moving wings continued. Some flocks circled the field and stopped, while others passed on. With the coming of night, the field would seem alive with the calls of the ducks and the rustling of hundreds of wings.

Now duck shooting has become so popular that the abandoned rice plantations of South Carolina often bring higher prices as game preserves than they did as rice plantations, when the industry was enjoying its best days. I presume the reason why the planters before the Civil War did not care for duck shooting was that these birds were then so plentiful in their fields, and hence so easily killed, that their killing was not considered good sport. It was looked upon more as "pot-hunting" than real sport.

When the ducks came in the fall in those days, they not only came in great numbers, but they stayed in the fields day and night, for then it was the practice of the planters to flood their fields as soon as the crop was harvested and keep them flooded until late in the winter, when work for another crop had to be begun. When there was a late fall, from the rice stubble a second crop would grow and mature small heads of rice, so that these, together with the shattered rice from the first crop, afforded an abundance of food for the ducks. There was no need for them to go anywhere else. Early in November they began to pour into the fields in large flocks, and, not being constantly shot at as they are now, they re-

mained until early spring. On every rice plantation, there
was some one Negro who was known as the plantation duck
hunter, and this was his only work.

Nathaniel Heyward's duck hunter was named Matthias,
and if all the stories I have heard of him are true, he must
have been a good one. Certainly his reputation long outlived
him.

Some of the older Negroes of my rice-planting days remembered Matthias, and used to talk about him and his
heavy long-barreled musket, whose deep roar could be heard
for miles on an early morning, and which could scatter shot,
and plenty of it, "ober a quarter-tas," as the Negroes expressed it. From what they said, Matthias must have been
a good man even to carry his musket, and a brave one to pull
its trigger, for I imagine those old muskets were the "Big
Berthas" of the sporting artillery of those days, and with
none of the scientific devices in use today to absorb their recoil. Matthias received the full rearward impact when he
pressed his finger on the trigger.

"Enty you 'member Mytias' shouldur, how 'e right shouldur stan' back monuh den 'e lef'? Dat muskit do um; 'e
kick wusser den eny mule."

Like other plantation duck hunters, Matthias, every morning during the season except Sunday, while the stars were
yet shining, would quietly paddle his batteau down the river
and noiselessly fasten it with a rope—no noisy chain for him
—to a stake at some landing place in the marsh. He would
then sit on the river bank until it became light enough for
him to see, when he would quietly stand up and survey the

field until he thought he had located, by the noise they made, the largest flock of mallard ducks feeding in the stubble.

Immediately then Matthias got busy. He quietly followed the river bank, musket in hand, until he came to a check-bank through the field, which he knew would lead him to the flock he had decided upon. He walked the check-bank for some distance, regardless of the heavy dew upon the tall grass and weeds. Cautiously and very slowly he moved, stooping very low and stopping every now and then to listen. In his right hand he carried his musket trailing by his side, and quietly he slipped it through the weeds so as not to shake them. When he came within about two hundred yards of where the ducks were feeding, he got down on his hands and knees and later on his stomach, and crawled along, trying to hide behind the dead grass and weeds on the side of the bank. Every now and then he stopped and, slowly raising himself on his left arm, peered through the weeds to see how the ducks were behaving.

The ducks in the meantime, unconscious of impending danger, continued to feed. They would thrust their heads down in the shallow water until only their tails could be seen, and there was a great quacking going on all the time; at one moment it sounded as if they were greatly enjoying themselves, quacking from pure satisfaction; then there would be heard a note of alarm. These notes of alarm always caught the ear of Matthias, for he knew the ducks had sentinels on the lookout. Whenever a drake flew up in the air a few feet and poised on flapping wings as he cast a keen and practiced eye over the field, then it was that Matthias instantly stopped and lay flat, not moving a muscle. Giving the duck time to alight, he would slowly crawl on again, repeating this whenever a duck flew up. Finally, when he reached the exact spot he wanted, still lying flat he cautiously moved his gun forward, placing its butt firmly against his shoulder. Then, digging his toes in the bank to brace himself, he aimed quickly into the middle of the flock and pulled the trigger. A tremendous explosion followed, which on the morning air carried for miles.

A foot or two back from where he had fired, lay Matthias, but only for a moment. Quickly he recovered from the recoil and stood up. He could see the flock flying fast out of the square, but he saw, too, many dead ducks lying on the water with their heads under it as if still feeding, but with their tails drooping. Others, wounded, fluttered and splashed here and there, beating the water with their wings, sometimes diving, trying to escape.

Leaving his musket on the bank, Matthias dashed into the water. Through the face ditch he went, and was soon wading among the dead and wounded ducks, filling a large sack, which he had brought with him. At about the same time, long jarring detonations of the heavy-loaded muskets of Negro duck hunters could be heard on other plantations, up and down the river, sounding like quarries blasting rock, and sometimes puffs of smoke could be seen as the sun reddened the eastern sky, against which flew, in widening circles, black clouds of startled ducks. Matthias's day's work was done.

Only once did Matthias ever fall from grace. One year for a while he failed to bring home as many ducks as the overseer thought he should, though he was furnished with his usual supply of powder and shot and was heard to fire his musket as often. The overseer, to teach him a lesson, took away his musket and told the driver to put him to work with the other Negroes digging a canal. A few days later, the overseer rode to where the canal was being dug and saw Matthias, who looked at him very pleadingly.

"Matthias," Mr. Thomas asked, "do you think you can kill ducks now?"

"Obshur, jus' try me, fur de Lord's sake, try me!" implored Matthias.

And Mr. Thomas gave him another chance.

15

THE BEGINNING OF THE CIVIL WAR

BEAUFORT is, with the exception of Charleston, the oldest town in South Carolina. A most attractive little place it was in 1861 with its warm sunshine and its fresh soft breezes from the ocean. Live oaks shaded its white, shell-paved streets, along which stood the homes of planters, with broad piazzas facing south. The town is on a bluff fourteen miles from the ocean in a bend of the Beaufort River. The harbor at the mouth of this stream is naturally perhaps the finest on the South Atlantic coast.

Beaufort has had an eventful past, but I shall refer to but one episode in its history—its capture in the autumn of 1861 by the Federal navy. Unlike Charleston, which in all its history has never been taken from the sea, Beaufort, without forts or fortifications, fell an easy prey to the Northern ships of war. Shortly after its capture, Federal troops took possession of the town and retained it until the flag of the Confederacy had been furled from Virginia to the Mexican border. In fact, some of the invaders were so captivated by the place that they decided to remain there. Their descendants are in Beaufort today, and are among its most useful citizens.

The Federal troops received a warm welcome from the

Negro population. In those days there were many slaves, Gullah Negroes nearly all of them, in Beaufort and on its adjacent sea-islands. These slaves were told that the Northern troops had come as their deliverers. Many of them, believing it, hastened to fraternize with the soldiers, who at first could not understand the strange dialect they spoke. A number of these Negroes even joined the Union army.

And how these slaves, still legally the property of their owners, delighted to swagger around in their new uniforms, with light blue trousers and dark blue coats, with their little forage caps jauntily set on the side of their heads! What a change it must have been for these Beaufort Negroes to quit sweeping the sidewalks along the shell-paved streets and to parade up and down them with their muskets at "shoulder arms," trying to keep step to the beating of drums! The more noise the drums made, the more they strutted in line until they strained to the uttermost their brass-buckled belts.

Slaves in Beaufort in those days seemed to associate noise with freedom—the more noise they could make, the freer they felt. Not long after the war, a Negro boy on my plantation, whose father had been one of the soldiers in Beaufort, stood one day in front of the office and persistently blew a tin horn. Finally the overseer could stand it no longer and called to him to either stop blowing the horn, or move on. "Wudduh use fuh be free," the boy answered, "ef can't blow tin horn?"

There must have been a great deal of noise in Beaufort during the war, with the Gullah Negroes marching and counter-marching through its streets, and drums continually beating, while little darkeys standing on the sidewalks no doubt blew tin horns to their heart's content.

During the war the South was dependent entirely upon its own production of food crops for the support of its civil population, as well as for its soldiers in the field; and agriculture in South Carolina, as in other Southern states, was carried on largely by slave labor. It was most natural, therefore, that the Confederate government preferred that the slaves in and around Beaufort should be plowing instead of parading. It soon required that all of the slaves in a radius of twenty-five miles of Beaufort should be removed to the interior of the state, and there employed in the growing of crops.

The Combahee River, which separates Beaufort County from Colleton County on the north, was bordered for miles in those days by large rice plantations. The slaves on these plantations had heard of the war. On that April day when Fort Sumter fell, and when they were planting rice in the fields, they could faintly hear the sound of distant guns in Charleston Harbor, but as to what it was all about, their ideas probably were very hazy. If they thought their freedom was in any way involved in the issue which that day had been joined between the North and the South, they gave no sign, but continued at their tasks.

Charles Heyward's plantations lay on the Colleton side of the Combahee, across the swamp and the river from Beaufort. When the Federal troops took possession of Beaufort, his rice crop was in process of being harvested, but, as he says in his diary, "It was not all threshed out in consequence of the war and the want of transportation to market." Still conditions were not so disturbing that they prevented him from attempting to plant the next year.

From his diary it appears that he succeeded in planting the crop in the spring of 1862, but soon had to abandon it, principally on account of the requirement of the Confederate government that the slaves be moved away. In June of that year, he says in his diary: "In March, fifteen of my Negroes, including three women and one child, left the plantation and went over to the enemy on the islands. . . . Later in the spring moved to Wateree all the people, except a watchman on each place, and some of the old people who did not want to leave." Thus it came about that my grandfather's Negroes went from Combahee to Goodwill, the plantation of my father, Edward Barnwell Heyward, on the Wateree River, twenty-two miles from Columbia, in the center of the state.

Only a few old Negroes, as my grandfather says, remained on his Combahee plantations, and he made a record of their names. Until the close of the war, each week one of his overseers, an old man, drove from his home in the pineland and distributed rations to these Negroes, who had nothing to do but fish in the trunk docks or doze peacefully in the sunshine in front of their houses. These old Negroes did not bother themselves about "de Nyankees een Buefu't," for how could freedom improve their lot? Were they not being taken care of, although scarcely one of them had struck a lick of work for years?

When my grandfather's Negroes left Combahee for Goodwill, it was the first time any of them had been more than a few miles away from the plantations where they had always lived. I have wondered why those slaves went; there was no power which could have been evoked to compel them to go. Had they wished to, any night they could have crossed the Combahee River, walked to Beaufort, and been practically free. Why didn't they? Simply because they didn't "trus'," as they would have expressed it. "Me know old Maussuh; me ent know nutt'n 'bout dem Nyankee." I am sure this was the deterring thought in their minds, especially with the older slaves and those who were leaders among them. All of these Negroes went to the Wateree in charge of one white man, Squire Jones, one of my grandfather's overseers.

Squire Jones used to tell me about that trip to the Wateree with all those Negroes. As I think of that kindly hearted little man, with his blue eyes and his snow-white hair and beard, it seems almost a miracle that he managed them as he did. He told me that all he had to do was to tell March, the head driver at Rose Hill, John, at Pleasant Hill, Ishmael, then seventy-four, at Amsterdam, and Cornelius, the driver at Lewisburg plantation, to tell all their people on a certain day and hour to assemble at the "weather house" on the Charleston and Savannah Railroad. His order, Mr. Jones said, was carried out without the slightest protest or trouble. The Negroes bundled up their belongings and were all at the appointed place long before the hour of the arrival of the train. Some carried their chickens with them in baskets covered with cloth, and some wanted to carry a pig or two, but this was not allowed.

Arriving in Charleston, they traveled by the South Carolina and Georgia Railroad to Gadsden station, eleven miles from Goodwill, where my father's wagons met them and carried those who were for any cause unable to walk. Upon their arrival at Goodwill, they were given homes which he had prepared for them. It was three and a half years before they saw Combahee again. When they did return, they brought back with them pleasant recollections of their stay " 'een de up country," as they always spoke of it. A few young women never came back, for they married Wateree Negroes, and I can recall several men who brought back with them wives whom they found at Goodwill.

Goodwill, on the west of the Wateree River, was a plantation where cotton and corn were grown, the labor being done by eighty-two slaves which my father had bought in 1858. For years a small acreage in rice had been planted in the river swamp, which was drained into the river and watered from a large millpond whose water also operated a combination grist and saw mill. Upon the coming of the Negroes from Combahee, who knew nothing about the planting of cotton, it was decided to increase to a considerable extent the acreage in rice. The Negroes spent the latter part of 1862 in clearing and ditching more land in the swamp, and also in increasing the length and height of the river bank, for at such work they far surpassed the Wateree Negroes.

There was one thing, however, which the Low Country Negroes, accustomed to tide streams, never seemed able to understand, and that was why the river always flowed one way, instead of back and forth, as did the Combahee. "Enty de ribbuh gwine dry up? Huccome 'e kin run one way all de time?" were questions they often asked. Those Negroes must have felt at first very strange and somewhat timid in their new surroundings, especially at night, for they seemed to have been given to hearing sounds which they had never heard on the Combahee. One of these sounds was the loud striking of a clock.

In the coach house at Goodwill was a clock which a classmate of my father's at the South Carolina College had made for him. The Combahee Negroes, the older ones who had been on the Wateree, used to tell me how loud this clock struck, and how it frightened them as they wandered about in strange places at night.

"Me sway, Maussuh, wen me duh gwine home one daa'k night en git all tangle up en dem strange paat dat clock 'trike so loud oonuh kin yeddy um fuh mile. Me h'aat duh mos jump outa me mout. Me staa't fuh run en me nebbuh top till me git spang me own do'. Me berry glad you pa lef um tuh Wateree," they used to say. These tales of theirs aroused my interest in a clock which had such wonderful striking power, and I began to make inquiries. I found the clock had not been left at Goodwill after the war, but had been stored in Charleston. I looked it up and had it installed on my plantation, where much to the surprise of the older Negroes, it did not strike. Upon investigation, I found that it had never struck. Its striking at Wateree was all in their imagination. Yet, as I think about it now, I can hardly blame some of them for declaring the clock struck, for how many of us today can say positively whether the old clock which stood on the dining-room mantel when we were children struck or did not strike the hours which regulated our young lives?

During the three and a half years that the Combahee Negroes stayed at Goodwill, Mr. Jones was in charge of them. In his talks with me about those critical times, he said he had no more difficulty in managing these Negroes than he did before war was declared. The slaves were always ready to obey any order given them, and there was no visible change in their demeanor. The drivers saw to it that the work was properly done, and the routine of the plantation moved along as it had done on Combahee.

On Goodwill, as had been the case in the Low Country, the old Negroes and the children had nothing to do. Clarissa, the child nurse, probably had her hands pretty full, for the mothers worked on the Wateree, and the children had to be looked after.

I doubt if Doll, the sick nurse, had much to do, for the reason that the change to the sandhill section of the state must have benefited the slave children. Old Affy, the midwife, however, was kept busy. And the field hands found plenty of work of the sort to which they had been accustomed.

In February, 1865, a great change came. The South's system of slavery began to crumble.

General Sherman had finished his "march to the sea." He had taken Savannah, and his columns were entering South Carolina. Columbia was his objective, and the capital of the state soon lay in ashes. Borne by a strong westerly wind, the smoke from its burning buildings floated over Wateree like a great black cloud. Still there was no excitement among the Negroes. They were as silent as they had been in April, 1861, when they heard from a distance the opening guns of the war. March and Ishmael each evening received their instructions for the next day. The plantation work went silently on. Did those Negroes know that their freedom was so near? I cannot say, but, if they did, they said nothing, only patiently waited to see what would come.

Long before then, President Lincoln's Emancipation Proclamation had been issued, but the slaves in the swamps of
the Wateree knew nothing of it. Evidently the remarkable
scheme of General Benjamin F. Butler had failed. I refer
to the telegram he had sent some months previously from
"In the field in North Carolina," in which he made a request
to headquarters for "10,000 yards of strong kite string, and
also all the President's Proclamations there are in the office." [1]
One would naturally suppose from such a request that the
General proposed to fly a number of kites with the proclamation attached, and thus distribute them among the Negroes.

What happened to General Butler's kites I do not know.
Perhaps he never tried them, or maybe the strings were not
strong enough, or the proclamations became detached when
the kites rose. At any rate, none of the kites flew over the
Wateree, and none of the proclamations fluttered down. Not
until Sherman marched into North Carolina and Lee surrendered his army did the Negroes at Goodwill know they
were free.

Despite President Lincoln's proclamation, Charles Heyward did not fail to make his semiannual distribution of
clothing to his Negroes, although he most certainly must
have foreseen that the Confederacy could not hold out much
longer and that his slaves would be freed. Less than two
months before Sherman captured Savannah, and less than
four months before Lee surrendered, he made his last distribution, December 17, 1864, although he had great difficulty in obtaining cloth and had to pay very high prices for

[1] *The War of the Rebellion: A Compilation of the Official Records of the Union
and Confederate Armies* (Washington, 1880-1901).

it. On the day the cloth was distributed, he made this notation: "There being no woollen cloth to be procured, had to
give cotton osnabrugs to the Negroes. Those with figures received theirs at Wateree, those blank, at Combahee."

It must be said of this Southern slaveholder that with
ruin and poverty staring him in the face he still sought to do
his duty by his slaves, although he fully realized that soon
they would be his no longer. Nor did he forget the old
Negroes he had left on the Combahee, more than a hundred
miles away, whom he knew he would never see again, for his
health was rapidly failing. He still had those old Negroes
in mind, and his overseer had never failed to drive from
the pineland to give them their rations.

16

REFUGING AT "GOODWILL"

ABOUT the middle of April, 1865, my father drove from his Goodwill plantation to Columbia and returned with a Federal officer with him in his buggy. The Negroes were all called together, and the officer, standing in the buggy in their midst, made them a speech. During the speech he told them they were free and could come and go as they pleased.

I have often wondered how the Negroes received this announcement. I have asked Mr. Jones particularly what happened, and also a brother of mine who was old enough to remember. Both told me the Negroes did nothing. There was no demonstration on their part, according to these two eye-witnesses. All they remembered was that the Negroes loafed the rest of the day, talking among themselves and probably speculating as to when they would return to Combahee. The next day, Mr. Jones persuaded them to return to work, on the promise of my father and grandfather that the matter of their remuneration would be arranged. The crop had been planted, and it was to the interest of all that it should be cultivated and harvested. In fact, it was all that stood between both whites and blacks, and starvation.

A short time after this, my father, in consultation with the Freedmen's Bureau in Columbia, which had been recently established by the Federal government to look after the interests of the emancipated slaves, agreed upon a plan to work out the problem of caring for his growing crop. This plan, readily assented to by the Negroes and put into the form of an agreement, was signed by a Federal officer representing the Bureau and by both my father and grandfather, and to it were affixed the names of 246 field hands, of whom 181 were full hands, 40 were three-quarter hands, and 25 were half hands.

There are probably very few agreements of this nature in existence today and therefore this one is quoted in full:

Terms of Agreement between Charles and E. B. Heyward, Esqrs., and certain labourers.

"Goodwill" Plantation
Richland District, S. C.
5th June 1865.

We hereby bind ourselves to remunerate the services of all those negro labourers signed below, who shall have, at the expiration of the present year, faithfully, performed all the obligations of this contract, by sharing to them one third (1/3) of the present growing crop when gathered and housed; allotting one full share to each "whole hand," one three-quarters (¾) share to each three-quarters hand one half (½) share to each "half hand." To clothe, feed and house them as usual, to afford medical attendance and protection to them all including the old infirm and young, now on this plantation during the entire term of agreement of service.

We claim the right to punish in any lawful manner all cases of misconduct and of laziness, and in all cases of desertion and of insubordination to discharge the offender instantly from the plantation and to be relieved from that moment of all and every obligation of this agreement binding upon us.

We will endeavour to be just and kind in our treatment and to continue retaining the respect and confidence of these labourers, freedmen by order of the military authority of the United States Government, heretofore so obedient, industrious and faithful.

Chas. Heyward

E. B. Heyward

In presence of

F. W. Johnstone

We freedmen on plantation of E. B. Heyward, Esqr. signing below accept this offer and hereby bind ourselves by this contract of labour from this date till Janry 1st 1866, for Charles and E. B. Heyward Esqr, to conduct ourselves in an orderly and proper manner, to be prompt and faithful in all work set us, recognizing all lawful authority of our employers and their agents, and to conduct ourselves in such manner as to gain the good will of those to whom we must always look for protection.

Below the names of the Negroes this signature appears:

Michael Murray, Capt.

Co. E, 25th Ohio Infantry,

Headqrs B. B. F. & A

Dist. So. Car.

Seven months later the following statement was issued:

Columbia, S. C.

Jany 8th, 1866.

I certify that Mr. Charles Heyward and Mr. E. B. Heyward have faithfully performed all the obligations of this contract and are hereby finally and perfectly released from all demands of the freedmen, parties to this contract.

Ralph Ely,
Bt. Brig. Genl,
& A. A. Cmdr.

As soon as the crop was harvested and divided, the Low Country Negroes sold their portion of it at the high prices then prevailing and returned to Combahee, all in one body as they had left it. My father arranged for their transportation by train to Charleston, and from there he managed in some way to have them carried up the Combahee River in the "John Adams," a government gunboat. They were landed several miles from my grandfather's plantation, and, without any permission from him, each family took possession of its former home. For a year they waited to see what next would happen.

As my grandfather sat on the piazza of his house at the Wateree, his former slaves stopped on their way to the station to bid him goodbye. All they said was that they were going home, and would look for him soon. He never returned to Combahee and did not see them again. Broken in health and staggered by his losses, Charles Heyward could not recover under the final blow. The emancipated slave could look forward to a better day for himself and his descendants, but the old slaveholder's day was done. He soon went to his grave and his traditions and his troubles were buried with him.

On the agreement made by my grandfather and my father with their recently emancipated slaves, I notice that my father made certain notations after the names of some of the Negroes, stating that they had either died or "runaway," as he expressed it, between the date the agreement was signed and the time the crop was harvested and divided. According to these notations, two women had died and four had run away, while of the men fourteen had run away and none had died. In later years, I knew one of the women and several of the men.

I doubt very much if at the time my father marked these Negroes as having run away he suspected where they had gone, or that he would ever see them again. If I am correct in this surmise he had a surprise coming to him, for a year later when he returned to Combahee, he found every one of them there. They liked best the Low Country where they were born and reared, and as soon as they realized they were free, they went back to it, walking all the way.

Among the men who slipped away, I find the names of two who were later great favorites of mine, Mingo Powell and Fay Sheppard. At that time, Mingo was thirty-four years old and Fay only a boy of seventeen. Both of these Negroes had belonged to Amsterdam plantation and both were there when my father set foot on it again.

Mingo was a very small Negro, but an excellent trunk-minder. When through a slight mistake he would allow the water during the stretch-flow to get a fraction of an inch above or below the mark at which he had been instructed to keep it, he would always tell the truth and take the blame. He was so small and weighed so little that it was often hard for him to raise a trunk door, but finally he always managed to get it up. His death was a loss to the plantation, and I recall very distinctly going into his house to see him just before he died. He could not speak, and by his pulse I knew that he was nearly gone; still he was glad to see me.

If little Mingo walked all the way from Goodwill to Combahee with Fay Sheppard, and tried to keep up with him, he must have had a hard time getting there, for Fay was one of the finest specimens of humanity, white or black, I have ever seen.

During the period of the slave trade African tribes were often at war with each other, and occasionally the chief of the conquered tribe would be taken and sold into captivity. Often, as I looked at Fay, I have thought he must be the descendant of one of those chiefs. He stood at least six feet four inches tall, his body tapering from his broad shoulders to his waist, and he carried not an ounce of superfluous flesh. His features were far more Caucasian than African, and this was emphasized by a mustache and imperial. Fay's strength and endurance were wonderful. He could do three times more work in a day than any other Negro on the plantation. When only three years old, he was valued in Nathaniel Heyward's estate at a hundred dollars. Charles Heyward, had he lived, would have been lucky to have Fay on his plantation, for he was, in my opinion, the most valuable field hand ever owned by any of the Heyward family.

Fay had one fault, and one only. As he grew older, his fondness for whiskey increased, and, having purchased a horse and roadcart, he drank too much one night and when his horse jumped violently Fay fell backward out of the roadcart and broke his neck.

When in the spring of 1865 the slaves of Charles Heyward, with the thousands of other from Virginia to Texas, were freed, not only were they declared free, but on them was soon to be bestowed the full rights of citizenship. The South was placed in a terrible situation. The loss of billions of dollars could be overcome, but what her people dreaded most were the political and social problems which they knew would arise. A great black cloud hung like a pall over the once "Sunny South," a cloud which seemed to have no silver lining. Even her most optimistic citizens could not see much hope for the future, and yet they never gave way to despair, for the future of their descendants was at stake.

In the past, the people of the South have made mistakes, mistakes for which they have paid dearly, and the greatest of these was the adoption of the system of slavery. By its adoption, great agricultural development was made possible, but only a limited section was benefited, and that for only a little over a century. In the end, it retarded the development of their natural resources, proved a blight to their land, and a curse to their children.

While the North encouraged immigration and by means of it developed industries, the South clung to agriculture. Having had slavery thrust upon it, the South had to employ the slave, and he was fitted to labor only in the field and not in the factory. As a result the North prospered while the South lagged behind.

The South's political leaders, and especially South Carolina's, though many of them were men of great ability, did not, between 1835 and the outbreak of the Civil War, seem to grasp fully the trend of the times against the owning of slaves, and continued to champion that institution. For this position they were to a certain extent excusable, for Northern Abolitionists most bitterly denounced the Southern slave-holder, though many of their own ancestors had owned slaves, and for years many Northern slave ships had plied between the African coast and the ports of the South. As a result of this, intensely bitter feeling was aroused between the sections, which ended in a long and bloody war.

About five years ago, on a cold December afternoon, I drove from Columbia to Goodwill. I had not been on the place for some years, and I inquired whether there were still any old Negroes there who had been slaves. I was told there was only one, an old woman, named Betsy Anderson.

An idea struck me: I would have a talk with Betsy.

As I approached her house, I saw an old woman at a little distance, her arms filled with small sticks of firewood, which she had picked up in the woods. I noticed also that the little yard in front of her house was so thoroughly swept that its sandy soil looked as clean and as white as the sand on the seashore. Her waiting so late on a cold winter afternoon to gather firewood and the cleanliness of her yard reminded me of the Negroes on Combahee. I felt certain she was a Low Country Negro and that she had been one of my grand-father's slaves.

I knew, too, that if I were right in this, I should have an opportunity of checking up, on the very spot itself, all Squire Jones had told me of what happened at Goodwill on the day the Combahee Negroes were told of their emancipation.

I decided I would not tell Betsy who I was until I had

questioned her about people and events which she had probably not thought of for more than sixty years.

When the old woman came up to my car, I asked her name. With a low curtsey, she told me.

"Betsy, were you born here?" I asked.

Seeming to think for a moment, she said:

"No, sur. Me ben born in de Low Country."

"Whereabouts in the Low Country?" I inquired.

Again she seemed to be thinking before she answered. Maybe after so many years she had forgotten where she was born, or perhaps she was wondering why I, a stranger, had so much curiosity. Finally she said, "To Myrtle Grove."

"Why," I said, hoping to reassure her, "I know all about Myrtle Grove."

"Oonuh yeddy 'bout Murtle Grove? Huccome?" The old woman was so surprised and excited that she had lapsed into the Gullah dialect of her youth.

That word, "huccome," and the way she looked at me when she said it, meant that she was inclined to doubt my statement. I had to convince her.

"Do you remember Charles Doiley, the carpenter at Myrtle Grove?" I asked.

Immediately her whole expression changed from one of distrust to one of surprise. "Him bin me cuz'n," she exclaimed, and then a thought seemed to strike her. Letting fall her armful of firewood, she drew nearer, and, looking me in the face, asked: "Mistuh, who you bin?" "Never mind who I am," I told her. "Do you remember Charles Heyward?" "Great Lawd! Him bin me Maussuh." Then I told her who I was.

We had a long talk, during which I asked her a number

of questions. All the while it seemed to me I was talking to one from the dead, about the dead.

Then I asked Betsy if she remembered the day my father drove back from Columbia with a Yankee officer in a blue uniform with him in his buggy, and how, when March and Ishmael called all the Negroes together, the officer had made them a speech and told them they were free.

"Yaas, suh, me members um well. Dey all yedduh 'roun de buggy, en 'e talk 'um jus' like you say. De crowd bin stan' jus' weh you see dat tree. Me bin right 'mong dem."

"What did the Negroes do, Betsy, when the officer got through talking to them?"

"Why, suh, dem 'ent done nutt'n!" The very reply Mr. Jones had made to this question.

"They didn't holler and shout and have a big time?"

"No, suh. Me nebber yeddy dem say nutt'n."

"Did Squire Jones order Ishmael to tell them all to go back to their work?"

" 'E mus' hab tell um. Me know dem al gone wuk en finish de crop; en den dem all pick up en gone back Combee."

"Why didn't you go?" I asked her.

"Me marry Henry Anderson. Him blongs here, en me couldn't lef' um."

When I got home that night I looked through the old slave record book. I found this entry:

Name	Age	Occupation	Residence
Betsy	24	Field hand	Myrtle Grove

Betsy died three years ago at the age of ninety-one. I liked to talk to the old woman, for links connected us which went far back into the past.

17

CHANGED CONDITIONS ON THE COMBAHEE

MY FATHER, Edward Barnwell Heyward, the eldest son of Charles Heyward and Emma Barnwell, was born in Beaufort, his mother's home, May 4, 1826. As a boy he lived during the winter at Rose Hill, spending his summers in Charleston, where he went to school and later attended the South Carolina College in Columbia, from which institution he was graduated in the class of 1845. A favorite grandson of Nathaniel Heyward, he was enabled after leaving college to make an extended trip to Europe, during which he cultivated a fondness for art and literature.

Upon his return, he married Lucy Izard of Columbia, a descendant of Ralph Izard, one of the first two United States senators from South Carolina, and made his home in Charleston. It was some years later, on account of his wife's declining health, that he purchased Goodwill plantation. Several years after her death he married Catherine Maria Clinch, the daughter of Duncan L. Clinch of Georgia, a general of the United States Army and later a rice planter. Thus it came about that I inherited from both sides of my family my desire to plant rice.

My mother's family had settled in southeast Georgia,

where several became rice planters, among them my mother's grandfather, John Houstoun McIntosh, whose plantation, The Refuge, lay on the Satilla River near Brunswick. This plantation remained in the possession of his descendants and was planted by them until 1906. There was no better or more profitably conducted rice plantation on the South Atlantic seaboard than The Refuge. Its lands were not only exceedingly fertile, but it had another great advantage, for it was situated where the Georgia coast curves inland, and tropical storms often passed it by. When its crops failed to yield an average of sixty bushels per acre, its owners were disappointed. It might be added that The Refuge was the last large rice plantation to be planted in Georgia.

When my father bought Goodwill, he had in mind, principally, the planting of cotton. From the large millpond on the place, however, its former owner, a gentleman from the Low Country, had been able to irrigate some of the land bordering the river and to plant a small acreage in rice, and my father proceeded to follow his example.

When the Civil War began, Charles Heyward, as already told, was forced to move his Negroes from his Combahee plantation, and his son was glad to be able to offer him Goodwill on which to locate them. For a short time he remained there to assist his father and Mr. Jones in getting the Combahee Negroes settled, in order that not only cotton but food crops might be grown for the benefit of the Confederacy. As soon as this had been accomplished, he applied for and received a commission as lieutenant of engineers in the Confederate Army. As such, he served along the coastal section of South Carolina until the close of the war. When he returned to Goodwill, he found the plantation affairs proceeding as smoothly as when he had left there. Squire Jones and the Combahee Negroes were glad to see him back, for they had known him for years.

While serving with his company in the Low Country, he had occasionally been near enough to his father's plantations on the Combahee to visit them. Whenever he did so, he wrote his father most interesting reports, telling him about conditions there and about the few old Negroes who had been left behind when the others went to Goodwill. The plantations, my father said, looked much the same except that there were so few Negroes to be seen and no work being done. He reported that the Negroes were being regularly supplied with food and seemed well and contented. The silence which prevailed everywhere, judging from his letters, must have gotten on his nerves, and I can well appreciate this, for fifty years later, when rice planting was ended on my plantations, that same silence certainly got on my nerves. Of course, the reason my father wrote my grandfather about his plantations whenever he had an opportunity to do so was that he knew Charles Heyward's heart was yearning for Combahee, which he was destined never to see again.

Barnwell Heyward spent the year 1866 at Goodwill, which must have seemed very much deserted after the departure of the Combahee Negroes. Many responsibilities rested on him: his father was ill and passed away during the year; he knew nothing of conditions in the Low Country; and the Negroes upon their return had promptly taken possession of the plantations, and it was a question whether the freedmen might not be allowed by the government to retain these lands. Were Rose Hill and Pleasant Hill, Amsterdam and Lewisburg to be included in the lands said to be promised to the freedmen—to each Negro forty acres and a mule?

Conditions, too, in Richland County, in which Goodwill lay, were greatly disturbed. Columbia was still occupied by Federal troops, and throughout the county there was great unrest among the Negro population. With the consent of the military authorities, a local company of young white men, of which my father was made captain, was organized in the neighborhood of Goodwill. The purpose of this company was to suppress any conflicts which might occur between whites and blacks; to reassure the whites of safety; and also to protect the blacks should such a necessity arise. This purpose Barnwell Heyward set forth in a long letter to General Wade Hampton, who had, at the close of the war, returned to his home near Columbia. In this letter there was also a request that the General come to that section of the county and make an address to both races. The Negro population of the state, regardless of the fact that he had been a large slaveholder, had great confidence in Hampton, which they showed ten years later when many of them voted to elect him governor, and in addition to this he was the idol of the whites. Barnwell Heyward felt that Hampton could do more than anyone else to pour oil on the troubled waters of those critical times.

When Charles Heyward died, he bequeathed his four plantations on the Combahee to his son Barnwell and his daughter Elizabeth, the wife of General James H. Trapier of the Confederate Army, and he stipulated that they should be divided equally between them. My father insisted that his sister should have the first choice, and she chose Rose Hill and Pleasant Hill, leaving her brother Amsterdam and Lewisburg, which together contained eight thousand acres of rice lands and fifteen hundred acres of adjoining highlands. When compared with the number of acres planted by his father and grandfather, these two plantations must have seemed to my father very small indeed, but without a dollar of his own, and being unable to obtain any advances in Charleston, to undertake planting them was almost hopeless.

After many futile efforts, he finally succeeded, through the assistance of a first cousin of his wife's, Mr. John Kilgore, a man of means and a resident of Cincinnati, Ohio. Mr. Kilgore agreed to endorse his notes for $15,000 and arranged with a wholesale grocery firm of Cincinnati to advance him plantation supplies to that amount, at various times during the year.

This old contract between the Cincinnati house and my father is today most interesting, for it shows that in those days Southern planters had to rely upon themselves to obtain financial assistance, and it shows also the high cost they paid for it. This contract of Barnwell Heyward's, which now lies before me, stipulated: First, that the wholesale grocery firm should receive two and a half per cent commission upon all groceries purchased and shipped to E. Barnwell Heyward, and also interest on the money invested in these at the rate of twelve per cent per annum. Second, that Barnwell Heyward bind himself to ship to the grocery firm his entire crop of rice, he paying the cost of freight, storage, insurance, etc., and in addition upon the gross sales of the rice to pay a commission of two and a half per cent.

He was forced to continue this arrangement for several

years, but, even so, he was more fortunate than many rice planters of that day. I have heard that a first cousin of my father's paid as high as fifty per cent for the money he borrowed for his first rice crop after the close of the war.

In 1866 my father, after gathering his crop on Goodwill, and having made an arrangement to have the place kept under cultivation the next year, late in the fall moved with his family to Charleston. To that city he took with him several of his old family servants and also their families, many more than he should have taken for he had no certainty of being able to support his family, much less a yard full of Negroes. I doubt whether he ever thought of this. He was simply doing what many other former slaveholders of his day did. It was these loyal servants who had guarded their owners' families during the trying years of the war, and to desert them was beyond comprehension.

It was early in January, 1866, that my grandfather's Negroes had left Goodwill and returned to their former homes on the Combahee, and there, a year later, my father found them when he went to his plantations to begin planting under very different conditions from those under which his ancestors had planted for so many years.

Agencies of the government, during 1866, with the town of Beaufort as their headquarters, made futile efforts to help the Negroes in Beaufort and Colleton counties by giving rations to the most destitute among them, but the older Negroes on the Combahee were not able to walk to Beaufort, and the suffering and mortality among them greatly increased. Had my grandfather compiled a year later than he did his list of "Emancipated Slaves," he would have had to omit the names of a number of his old Negroes, especially of those who had remained on his plantation during the war and who had looked forward each week to the coming of the overseer from the pineland to give them their rations. In 1866 no overseer came. "Old Maussuh" had gone. Instead, "ole Stephney" (a Gullah word for hunger) came to take their places.

Late one afternoon in early January, 1867, after a ride of hours from Charleston on a much dilapidated railroad, my father arrived at Combahee and spent the night with a cousin whose plantation adjoined his. I can well understand how depressed he must have been at the changes he saw on every side. Dwelling houses, threshing mills, stables, barns, and all had been burned; only the Negro settlements remained. But what seemed to surprise and hurt him most of all was the changed attitude toward himself of the Negroes, who had so feelingly bade him goodbye when, only a year before, they had left his plantation on the Wateree.

In a letter to my mother, written shortly after getting to the plantation, he comments on the run-down condition of everything, and says: "But as to the human part of it; Oh! what a change. It would have killed my father and worries me more than I expected, or rather the condition of the Negroes on the place is worse than I expected. It is so very evident that they are disappointed at my coming here; they were in hopes of getting off again this year and having the place to themselves. They received me very coldly; in fact it was some time before they came out of their houses to speak to me. . . . They are as familiar as possible and surprise me in their newly acquired 'Beaufort manner.' They are constantly in Beaufort, quite too much for their good. I can see quite clearly from all around me that the job is a big one."

My father had before him not only a big job, but at first a most discouraging one. Alone for some time in the great silent swamp, scarcely ever seeing a white face, he lived in a small house formerly occupied by the overseer in the Amsterdam settlement, and in close contact with nearly two hundred Negroes. These were, at first, as he says, somewhat hostile in their attitude toward him, because disappointed at his return to his own property; yet his determination to solve the problems which confronted him and to plant his lands did not waver. Before his death he succeeded in his carefully thought-out plans by which to manage these Negroes. When, his constitution shattered by overwork and anxiety, he died a young man, after four years of great hardship, he had the reputation of having been not only a successful planter and leader in solving the South's social and economic problems arising out of the emancipation of the slaves, but also the most popular planter of his day with the Negroes on the Combahee. He had been thrown with these Negroes since his boyhood and for this reason soon gained their confidence and respect, with the result that he soon had more labor in his fields than he really should have employed.

I am quite sure my father would not have lived even as long as he did under such trying conditions, and certainly would not have gotten on as well with the Negroes, had it not been for his sense of humor. He managed to see the funny side of things. Even in the midst of his work and his problems, when anything happened which he thought amusing he would write home and tell about it. One day he was much amused by some little Negroes in the rice field, and he wrote me a letter which he asked my mother to read to me. I quote from this letter: "I am sorry to tell you that the little children still behave very badly here. Yesterday, I was in the rice field and took a gun and shot a marsh hen, which dropped in the rice and I could not get it. There were some little children there minding the birds, or rather making believe, for they play and cook cow peas in their pot on the dam. I told them to go in and look for the dead bird, and they jumped right in, naked, and swam across the canal and looked about. One didn't follow, and I said, 'Why don't you go?' and the child said, 'Why, Maussuh, I is a gal!' 'Well,' I asked, 'Can't you swim?' and she said, 'Oh, yes.' 'Well,' I said, 'Go ahead,' and away she jumped heels over head and swam like a duck; didn't mind getting her frock wet a bit; thought it was high fun. They found the bird and brought it to me."

Further on in the same letter, referring to the Negro children on the plantation, he says, "They are good little things when you treat them well and can help you very much, and they take to water like a young duck. I am afraid the snakes will bite them, but they tell me, 'Ef bad snake come we will kick um en dem mout' en t'row mud en dem eye.' I believe the snakes like these little children because they leave so much hominy and cow peas scattered about on the bank when they go home, for they have no spoons and eat their victuals with a stick or their fingers. I hope you will see them next winter and that they will amuse you." I did not see them until a good many winters later on.

One of my father's theories as to managing the Negroes

after their freedom was that the owner of the plantation should, himself, come into close personal contact with them while they were at work. The old plan of management, he felt, must be changed; the day of the driver had passed, and even the overseer should no longer be an intermediary between the owner of the plantation and his Negroes. He realized that the freedmen would naturally resent working under the guidance of the same drivers who had managed them in the days of slavery; it savored too much of old times, although after the war the word driver was never used, the name foreman being substituted. Though my father employed a foreman, he never selected for this position a Negro who had been a driver before the war. Had he turned loose old "Wasp," for instance, on the plantation, I am quite sure he would have had few Negroes in his fields. But how Wasp would have enjoyed it!

I have always thought the word "driver," which was used all during the years of slavery, was a most unfortunate one. It came into use in the early days when the African slaves were first brought to Carolina in any number. As the Negroes, upon their arrival, could not of course understand or speak one word of English, it was quite a problem to manage them and teach them to work. The method almost universally adopted was to divide them into squads of ten and to place each squad in charge of a domesticated Negro, whose duty it was to stay with them day and night. When this Negro put them to work in the field, he was required to follow them foot to foot, so as to direct them how to use the few implements given them. In fact, this was the only way they could have been taught to use the hoe or the hand sickle. His walking behind them naturally suggested the idea that he was driving them, and hence the term came into general use and continued to be used for upwards of two centuries. It may, however, have originated with the Negroes themselves, for when they began to use the Gullah dialect such an expression as "oonuh dribe we too fas' " was enough to have suggested it.

After an illness of more than a year, brought on by exposure during the war in the swamps of the Low Country, and by worry after the war over the agricultural and racial problems confronting him, Barnwell Heyward was still firm in his determination, with God's help, to play his part in the solution of these problems. But all of them combined were more than he, in his physical condition, could contend with.

While the future often may have looked dark to him, I am quite sure it brightened when on January 26, 1871, he passed away and was laid to rest beside Lucy Izard, his first wife, and next to his father, Charles Heyward, in the church yard of Trinity Church in Columbia.

It was said of him, by a younger cousin, who for two years worked for him on his plantations and was closely associated with him, and for whom he had a great affection, "Barnwell Heyward was one of those of whom it may be said—'He died too soon for the good of his country' because his various talents, both practical and ornamental, were forever ablaze

'Like a lane of beams athwart the sea
Through all the circle of the golden years.' "

After my father's death his two plantations, Amsterdam and Lewisburg, became the property of his three sons, all minors, and were managed, until I became of age, by his brother-in-law, the brother of his first wife, Colonel Allen Cadwaller Izard, a graduate of Annapolis and a lieutenant colonel in the Confederate Army. For eighteen years Colonel Izard planted these places most successfully, paid off the money which my father had to borrow in order to get his plantations under cultivation, and furthermore was able to accumulate some money for the estate.

The Colonel was an excellent manager of rice-field Negroes and very rarely became angry with them. When he did, he knew just what to say, having served in the Navy. I cannot recall his getting really mad but once. Monday, old Hannah Robinson's son, as black as a crow, was the house boy on the plantation and waited on the table. The kitchen was about fifteen feet from the house, and one morning, starting from it with twelve biscuits on a plate, Monday ate eight of the twelve before he reached the dining room. How mad the Colonel was that morning, and also Squire Jones!

18

THE GULLAH NEGRO

FULLY ninety per cent of the Negroes on Combahee plantations were Gullah Negroes, and this was the case on every rice plantation in South Carolina and Georgia, as well as on the sea-islands of these states, where long-staple cotton was grown for many years.

A most kindly feeling existed during the years of slavery between most slaveholders and these Negroes, a feeling which was handed down to their descendants, even including the field hands who seldom came in contact with their owners. Of this I have had many evidences, some of which I shall never forget, for they touched me deeply, especially when the Negroes on my plantations and I came to the parting of our ways. As an instance, I recall a conversation I once had with old Judy Simmons, a field hand all her life, who had belonged to my great-grandfather and my grandfather. Judy was a typical old Gullah Negro, emotional and good-hearted.

My closest friend at Cheltenham Academy, a preparatory school near Philadelphia which I attended, was Isaac R. Davis of an old Quaker family. His home was near the school. Before the Civil War his family were strong Abolitionists. Many an argument he and I had over the question of slavery, about which neither of us knew anything except by hearsay.

In later years, Davis visited me on Combahee and I was glad to see that time had somewhat changed his ideas of how the slaves had been treated. He showed much interest in the Negro settlements where the slaves had lived and in the old Negroes who had themselves been slaves.

One day, while driving with him in my buggy over the plantation, I thought I would give him a little surprise, which would open his eyes somewhat as to how the former slaves regarded the descendants of their owners. Before us in the road I saw the short, stout figure of Judy, and when we had overtaken her I stopped the buggy.

"Judy," I asked her, "did you ever see a Yankee?"

"No, Maussuh," she said, "me neber see Yankee, but me yeddy 'bout um."

"Well, this gentleman is a Yankee," I said.

She looked at him as if he were some kind of wild animal and exclaimed, "Is *dat* a Yankee?"

I explained to Judy that the Yankee and I had been friends since we were boys, that he was a real good Yankee, and that he had always been interested in colored people. I added that he had come all the way from his home in the North to get her to go back with him. I told her he would give her all she could eat and wear, and would take good care of her as long as she lived. Then I asked her if she would go with him.

Looking at me with surprise and a rather hurt expression, she exclaimed, "Enty oonuh know me ent gwine lef' me Maussuh fuh go lib wid no Yankee? Not me!"

I had the pleasure of poking Davis in the ribs as we drove on.

A short time ago, I happened to come across the plantation time-book I had kept when I first began planting, and began also my association with the Gullah Negroes of our coastal section. As I ran my eyes down the list of names grouped daily under such headings as "plowing," "ditching," "harrowing," "hoeing," I was surprised to find how distinctly I recalled nearly every Negro whose name I had so often written. Their names and faces not only came back to me as if I had seen them yesterday, but I recalled incidents connected with many of them.

It has been said that to some white people all Negroes look alike, but when one is thrown with that race as constantly as I was for many years, he must necessarily become very familiar not only with their names and faces but also with the natures and ways of many individuals among them. I used to try to learn the ways of these Negroes, but I could never divest myself of the suspicion that they were learning my ways far faster than I was learning theirs.

Some years ago, Mr. A. Felix du Pont of Wilmington, Delaware, who now owns my plantations, came from his home to Combahee on a fine yacht. When the boat arrived, many of the Negroes came to the river to see it. I was standing on the bank talking to Mr. du Pont, and they all came up to shake hands with me.

A young woman, who had grown up since my day, held out her hand and said, "Boss, oonuh ent know me?"

"I don't know your name," I replied, "but you are September Singleton's daughter."

"My Lawd, Boss," she exclaimed, "how oonuh know dat? Pa dead too long!"

Replying in her own Gullah dialect, I said, "Gal, enty you fabor yo' pa? Enty you en him all-two stan' same fashi'n?"

She went on her way highly pleased. How I happened to know her was that the long departed September had been exceedingly ugly, and as his daughter was, too, and as their ugliness was of the same type, I recalled him as soon as I looked at her. I did not explain to her, however.

I venture to predict that within a very few years, little will be known of the Gullah Negroes of our former rice and sea-island cotton plantations. The reason for this is that these Negroes cannot now find work in the Low Country of this state, and consequently they are having to go elsewhere. They began to move away nearly thirty years ago when the acreage in rice and sea-island cotton commenced decreasing, and their exodus has continued in increasing numbers each year. When they leave the lands where their ancestors have lived, especially the younger ones, they quickly learn to drop their Gullah dialect. Some still remain where they were born and reared, but closer contact with the few whites in that part of the state will in time result in even these dropping their dialect. There is now little to keep the Gullah Negroes on our former rice plantations.

So much for the Gullah Negroes, considered as a whole. Let me now look through my old time-book and relate certain incidents concerning some of those whose names I wrote years ago. These may give some insight into the mental processes and natures of these Negroes as individuals who were fair representatives of the Combahee Negroes.

The first step toward the making of a rice crop was plowing the fields. In my time-book, the name I always wrote at the top of this list was that of Sammy Wright, the foreman of the plowmen, and also the "locus pastuh" of the plantation. Sammy was tall and broad-shouldered, with a good face and a pleasant smile, never surly, always good-natured. He and I were good friends and I sometimes liked to joke him, especially about his being a preacher, because he was by no means a saint. One day I said to him, "Sammy, I don't believe you could preach if I were there to hear you."

"Wuffuh?" he replied.

"Oh," I said, "you know that I know the life you lead and you have not the gall to get up in a pulpit and talk before me."

"All right, Boss," he said, "oonuh jus' come en try me."

I told him I would. Some night I would drop into his church and see what he had to say.

"Boss," he said, "oonuh come but don't come unexpected. Me wanta hab chance to pick a tex' dat will fit oonuh."

I have often wondered what text he would have selected had I given him the chance.

I see William Boggs's name in my book. He has been dead a number of years, but I remember him perfectly and also some of the amusing things he used to say. In stature he was quite small. Like Sammy, he was also a "locus pastuh," but of the second degree. When a large plantation adjoining mine was offered for sale, I jokingly asked William why he did not buy the place and plant it. He had been born and reared on it. Quick as a flash, he answered, "Boss, enty oonuh know w'en jaybud cy'an 'tay een 'e nes' no nuse fuh wee bud tuh en' tuk een!" This seemed to me pretty good business judgment on the part of a "locus pastuh."

There used to be an old saying among the Low Country Negroes: "De buckruh hab scheme, en de nigger hab trick, en ebery time de buckruh scheme once de nigger trick twice." I have never forgotten how once the Negroes applied this saying to me, and by all of them combining their tricks, they came near to frustrating my well-laid scheme.

I had a commissary on the plantation, but I myself knew nothing about running it, and the Negroes were fully aware of this. It required some experience to deal in a plantation store, crowded with Negroes, especially on a Saturday night. All of their purchases of groceries were made in five and ten cent amounts, the total of which both the clerk and the purchaser calculated in their heads. Even the little Negroes were very expert at these calculations. In order to expedite business, flour, sugar, meat, and so on, were put up in small packages, each kind in a separate compartment on the shelves.

One Saturday night during the planting season, both the clerk and the overseer had to be absent from the plantation. I told the clerk that it would be all right for him to go and that I would keep the store, if he could write on one of the packages of groceries in every compartment the price and also what the package contained, and would put the marked package in the front of the compartment.

Early in the afternoon, in some way, word got to the hands in the field planting rice that both clerk and overseer would be away. I happened to overhear one Negro call to another as their "covering harrows" passed, "Wuhruh we gwine do fuh eat tonight? De sto' gwine close."

I then did a very indiscreet thing. I told them that I was going to keep the store and they need not worry.

I had no sooner spoken than I realized I had made a grievous mistake. Telegraph, telephone, radio—I defy news to spread any faster than over the "grapevine" system of communication these Negroes operated. Immediately the surprising news was heralded for miles: "De Boss gwine keep sto' tonight!"

Knowing what was before me and hoping some would tire of waiting, I remained in the field until dark, and then rode slowly homeward. When I reached the store, it was just as I had feared. Hundreds of Negroes, it seemed to me, were awaiting my arrival. I tried to keep my nerve. Bidding them "good evening" in as cheerful a voice as I could assume, and hitching my horse to the rack, I opened the store and lit the lamps. The crowd immediately surged in. Relying on my marked packages and keeping my composure the best I could, I walked behind the counter and inquired what they wanted.

For a moment or two there was a dead silence. All looked surprised. Had they been underrating for so many years my ability as a store keeper? Then a young woman decided to try me. Pointing to a bolt of cloth on one of the very top shelves, she said in a weak voice: "Lemme see dat piece of clawt. How much for um?"

"No cloth tonight," I said. "It is too late, nothing but groceries."

They all got started then and everything went along swimmingly as long as I was careful to locate my marked packages. All the while I saw a surprised look on their faces. They had certainly been underrating me! Then I slipped up. Becoming careless, I handed out a marked package. Jim Burns, my youngest foreman, leaned over the counter and whispered to me, "Look out, Boss, you gi' um de sample package."

He had caught on to my scheme. The sample package was restored. I was directed where to place it. They were all very kind and considerate to me from then on.

And how I was amused one day watching Jack Snipe digging out a leak in my river bank. These leaks were often caused by large water rats making their nests deep down in the center of the bank. Jack was an old Negro. He had had smallpox and was badly pock-marked. In addition, he had lost the sight of one eye. Just as I came to where he was working, he struck a large nest of rats, and one after another, they began running out. Jack struck at every rat with his shovel, but, probably on account of his blind eye, he missed them. Finally, a very large rat, the patriarch of the nest and so fat it could scarcely run, made an effort to escape. Jack thought sure he had that rat, but again he missed. Exhausted from his efforts, he stood for a moment resting on the handle of his shovel, and I heard him say to himself:

"My Gawd! Dese rat dem all got 'way. Dey mus' be Mason."

Who would have thought that an old Gullah Negro would have heard of that old saying, "You can't hang a Mason"?

19

CUDJO MYERS AND TITUS WRIGHT

ONE old Negro especially stands out in my memory, Cudjo Myers, who, in appearance, behavior, and peculiar mentality, was more my ideal of the wild Africans brought to this country than any other Negro I have ever known. And he came rightly by his African looks and habits, for I have always been told that his father was brought direct from Africa. Cudjo, with his beardless old face, appeared silent and sullen to one who did not know him. He was much inclined to attend to his own affairs and not to mingle with the other Negroes. He never seemed to joke with them, nor they with him. If he had any sense of humor, I have never seen him attempt to indulge it. When he said or did amusing or startling things, it was not because he intended them as such; they were due more to his ignorance and unfamiliarity with everything outside of the rice field. The two predominating traits of this strange old Negro were his willingness to work and the paramount importance with which he invested everything entrusted to his care. The latter trait, however, was sometimes offset by the probability of his misunderstanding any order given him containing certain simple English words which he had never heard, and of whose meaning he had not the faintest conception.

Cudjo was born on Combahee in 1822 and began work in 1834 as a quarter hand for my great-grandfather. Twenty-seven years later, in the valuation of my ancestor's estate, none of the field hands was valued at a higher price than was Cudjo. When I first knew him he was sixty-six, and yet there was no better worker on the plantation. During all of his life, his work had been confined strictly to the spade and shovel, the hoe and reaphook. Never had he plowed a mule, nor had he attempted to hitch up a horse. Thus had Cudjo's life been spent.

At last the years began to tell, and work in the rice fields became too hard for him. He was made night watchman of the stables and commissary. It was then that my closer acquaintance with him began. Immediately upon being promoted to this, to him, most responsible position, he requested a gun. An old musket about seven feet long was given to him. Three times each night he loaded it with a very heavy charge of powder only. The first night he was on duty, everyone for miles around knew I had a new watchman, for at "fus daa'k," "middle night," and "day clean" Cudjo would emerge from the watchman's house and fire his musket, the tremendous report of which carried far on the still air. He called this a "warn," his notice that he was awake and about. Immediately after firing the "warn" he would retire to his house in the lot and sleep, I am sure, until the next "warn" was due, although he used to declare, "Me nebber shut me eye."

Awakened before daylight one very cold winter morning by a loud knocking at the back door, I opened it and found old Cudjo. He announced, "Uh terrible t'ing is come to pass; de puppy dog shut up in de sto' all night, en Mistuh Haskill 'table en 'e mule dun all de bu'n up!'"

Now the "puppy dog" was supposed to be under Cudjo's special charge, and that it should have been locked up accidentally in the store was, to his mind, a greater catastrophe than the destruction of my neighbor's stable and mules.

On another occasion Cudjo displayed this same trait. I had taken him to my home in Walterboro, the county seat, to work about the yard. Returning from a month's vacation during the summer, I found the old man, who was expecting me, walking back and forth in front of the house, evidently much excited. As soon as I approached, he exclaimed, "Did oonuh git de news? Me tell Mass Dick fuh write oonuh. Oonuh shoulda git 'um. De blue tuckrey hen dead."

The turkey hen, like the puppy, had been left in his care.

One day soon after Cudjo began watching the lot, old Squire Jones, being in a hurry to drive to his home in the pineland and seeing Cudjo near, told him to harness his horse to his roadcart, forgetting the old man's limitations. Coming out of the store a little later, the Squire was confronted by the sight of Cudjo struggling to hitch up the horse with its head facing toward the roadcart. Angry at the delay and at his own mistake at assigning such a task to Cudjo, Squire Jones began to berate him.

As I watched the two old men who had known each other for thirty-odd years, I had to laugh at the sight. The roadcart had been placed facing in the opposite direction from that in which Mr. Jones was going, and I could readily see how Cudjo had made his mistake. He had never thought whether a horse pulled or pushed a roadcart—hence he had turned the horse's head the way he knew Mr. Jones was going, despite the fact that this would cause Mr. Jones and the horse to face each other. The position of the roadcart had caused the whole trouble, and I thought Cudjo put up a strong defense. Pointing in the direction Mr. Jones was to go, he kept on mumbling, "Enty you bin gwine dat way? How kin oonuh go dat way ef oonuh hoss look dishyuh way?" Certainly an unanswerable argument!

One morning during a very hot summer Cudjo said to me in a most matter-of-fact way, "De cow duh hatch ten chicken."

I told him he was an old fool. His reply was, "Come see fuh yose'f, Maussuh. Me show oonuh."

He led the way to a cow's stall and there, very close to where the cow lay down at night, were ten puny looking little chickens, with every evidence of having been recently hatched.

I looked everywhere but could find no clucking hen, nor any hen which seemed to be in the slightest degree interested in the chickens. Despite these facts, I did not believe Cudjo's story until my doubts received a jolt. During a hot summer several years later, I read several statements in newspapers to the effect that in certain places throughout the country the heat had caused eggs to hatch. So, after all, maybe Cudjo was telling the truth, but gave the cow the whole credit when he should have shared it between her and the sun.

Cudjo was neither crazy nor stupid. He was simply ignorant. He had never had a chance.

Very few people could tell the ages of the Gullah Negroes, and scarcely any of the older ones knew their own

ages. On one occasion which I shall always remember I was
myself badly deceived; so much so that I have never forgotten it. I thought I knew the Gullah Negro, but one of
them certainly fooled me.

One day Colonel Griffith, the superintendent of the South
Carolina Penitentiary, called at my office and told me that
there were at the time in the penitentiary a number of old
convicts sent up for burglary, who had served twenty years
and who, he thought, should be pardoned, in view of the
fact that the law fixing their terms to life imprisonment had
been greatly ameliorated since they had been imprisoned.

Colonel Griffith was a just and most kind-hearted man,
and one absolutely to be relied upon, and I asked him if he
had in mind any particular one of those old convicts, for if
so I would begin by pardoning him immediately. He said
there was one old Negro by the name of Titus Wright, who
had been a most exemplary prisoner and a hard worker, having been convicted of a minor case of burglary and had been
in the penitentiary, as the record shows, for twenty years.

I had the pardon made out for Titus and handed it to the
superintendent, who went away from the office much pleased.
The next morning early Titus came to the governor's office
to thank me. He came shuffling in, somewhat out of breath
from the long walk, his first in twenty years. His hair was
white, his head evidently not having been shaved for some
time, and his face was deeply lined with many wrinkles. He
had exchanged his stripes for a cheap suit given him at the
penitentiary; he held an old felt hat in his hand and wore a
coarse pair of brogan shoes. He seemed very grateful indeed
for his pardon—a feeling I had discovered to be exceptional

among beneficiaries of executive clemency—and I could not help feeling very sorry for him. I asked him if I could do anything for him, for I feared he would find it very hard to make a living at his age, and after having been so long a prisoner. He did not seem, however, at all depressed or worried. On the contrary, it struck me he seemed quite hopeful, and when I inquired what he expected to do and where he was going, he said that he was going back to Combahee, where he had come from, and expected to live with me for the balance of his life. He asked me to write to my overseer and tell him to give him a house, for he was going straight there. Again he reiterated his thanks, bade me goodbye, and hoped he would see me again soon.

Some years passed. The governorship became with me a memory, and I was back on my rice plantation again. In the meantime I had never seen nor heard of Titus Wright; I had forgotten him and his pardon; had forgotten even the gibe of my overseer, who, when I asked him if Titus had come to the plantation on leaving Columbia, said, "Don't you know a nigger yet? That rascal 'larged' around White Hall for a day or two and then went to Savannah. I suppose he is in the Georgia Penitentiary long ago."

But I was to see Titus again.

On a hot summer day we were putting in a "river trunk," always a hard and anxious job. For hours I had stood in the hot sun, and my boots were muddy to their tops. The forty-odd Negroes digging the trunk dock were most of them both wet and muddy, and perspiration streamed from them as they rolled their wheelbarrows filled with mud up the slippery gang plank.

Early in the afternoon, my attention was attracted by a man walking towards us on the river bank. He was strolling leisurely along, and carrying, I could see, over his head a large umbrella. When he came nearer, I saw he was a Negro.

"Who de debil kin dat be?" I heard one of the Negroes ask. " 'E walk lukkuh 'e hab all day befo' um. Dat nigger mus' t'ink 'e own Combee!"

Finally the stranger approached. He stopped and hesitated for a while, for to get to me he had to leave the bank and I could see that he did not wish to soil his new tan shoes. He wore a new blue serge suit, a very bright red cravat, and a neat straw hat with a fancy band. The umbrella he continued to hold over his head.

"Who dat sporty nigger?" I heard another Negro ask. " 'E ent fur muddy, eh? I luk fur trow um en de trunk dock."

I paid no further attention until I heard a voice behind me say, "Howdy do, Mr. Heyward?" I turned.

"You don't seem to recognize me, sir."

"No," I said, "I don't know you."

"Why, Sir, I am Titus Wright."

"Titus Wright?"

"Yes, Sir, I am Titus Wright, who you pardoned."

I said, "The devil you are! Not that old Negro!"

"Yes, Sir, I sho' is Titus. How is Colonel Griffith?"

At first, I could not believe him; I told him to take off his hat. He did so and his head was clean shaved. I looked at his face. The wrinkles and furrows were gone. I had pardoned an old man—I could see him as he looked that morning in the governor's office—and now before me stood a man in the prime of life! We talked it over. He told me he was about twenty years old when he was sent to the penitentiary. He had worked hard, as the Colonel said. He had worried, too, a great deal, as he told me, and had become prematurely gray. Twenty years seemed a long time to him and to others in the penitentiary. He had been there so long that his associates and the guards thought he must be old and they spoke of him, as did the superintendent, as "Old Titus." He looked the part, too, as I could myself testify.

I asked him where he was living and how he was getting on. He said he lived over in Savannah and was making shoes; that he had only run over to Combahee to see his old country again. He also assured me that I had done right in pardoning him, for he had made, he said, a new start in life; had married a young woman, had two children, and was doing fine.

I wonder how Titus and his family are now. I really have a great curiosity to see him again. Has he reverted to the old man I first knew, or is he still growing younger? I wonder. But Colonel Griffith would never believe me about his "Old Titus Wright." He often asked me, "Are you sure it was Old Titus?"

20

THE FIELD HANDS

THE field hands, during and after the days of slavery, were the mainstays of every rice plantation, and these never numbered more than one-half of the total Negro population. The rest consisted of men with trades, women with special occupations, old people and children. All children were prospective field hands, and before they were old enough to work they knew the name of every square and the shortest way to it, for the first thing they were called upon to do was to carry dinner to their parents working in the fields.

Probably the earliest recollection of slaves born on a rice plantation in the Low Country of South Carolina was the whitewashed house, considerably larger than the Negro cabins and a little distance from them, where they used to spend the day playing with other children and were minded by a girl somewhat older than themselves. They probably recalled very distinctly the yard before the house—just how it looked, how clean and well swept it was, without a single blade of grass, and with the soil made hard by the constant tread of many feet, except where, here and there, they had dug it up to make playhouses. They would remember too, the Pride of India trees with clusters of yellow berries, under whose shade, lying full length on the ground, they slept when the sun was hot. There the old child nurse dozed, but with an eye open all the while.

To the child nurse would come nearly every morning during the winter a lady who would ask the nurse questions and talk to the children. She knew their names and showed an interest in them. The black children looked forward to her coming, and when they saw her their eyes would shine, for she often brought a basket and would give them good things to eat. Never were they afraid of her, for her face was kindly and her words were gentle, and she never scolded as the old nurse sometimes did. They hated to see "Missus" go, and, standing in a group before the house, would watch her as she took the path which led to the "sick house" standing by itself.

The necessity of having a place where the children of the field hands could be cared for while their parents were at work was early recognized on all well regulated rice plantations during the days of slavery. To the credit of the planters, it can be said that this was one of the first forms of welfare work put into practical operation in this country. At the nursery every morning the mothers left their children, and when their work was finished they returned for them. When they had nursing babies they were allowed to visit them during the day. A cook prepared food for the older children, and for those requiring it milk was sent from the plantation dairy every day.

One of the most intelligent and trustworthy Negro women on the plantation was always in charge of the nursery, and it was the general custom of the wives of the planters frequently to visit these nurseries, as well as the plantation hospitals, to see that the inmates of both were properly cared for. It was the rule to leave largely to the women of the planter's family the supervision of all welfare work. So much was this done that the charge has been made that the wives of Southern slaveholders, in seeking to meet the responsibility which they felt slavery devolved upon them, neglected their own children, leaving them too much to the care of Negro house servants, while they themselves looked after the condition of the children of their slaves; and there may be some foundation of truth in this charge.

The slave child's next recollection would be of the time when, a little older, he belonged to what the old drivers called "de shut-tail gang," and went with other children to the fields at dinner time with a cedar bucket on his arm, filled with "bittles" and would "progick" with each other on the way. The check-banks seemed so long that they were obliged to play as they went and to "debil" each other, the boys teasing the girls, trying to appear "bigity" and to dominate them. They would pull and shove the girls about and throw stubble and clods of dirt on them, trying to make them mad, and often succeeding in making them cry. Now and then a girl, a little more bumptious than the rest, would get thoroughly angry and standing up with arms akimbo, to a young tormentor, would out-talk him, whereupon the other boys would promptly interfere: "Boy, don't let dat gal rule ona, knock um." But at this point, discretion usually proved the better part of valor and it all ended in a laugh. On the way, they would sometimes eat a "leetle of de bittle" in the buckets, feeling quite sure that none would tell. And thus early there developed a tendency in the slaves to stand together.

At twelve years of age all children were put to work, but when a child was not strong or well enough grown, records show that the age at which he was made to work was advanced, and always the nature of the work, as well as the amount, was determined by the strength of the child. A child beginning work was assigned a task next to that of one of his parents, usually his mother, who was supposed to teach him how to do the work properly. Should the child finish his task sooner than his parents did theirs, he was allowed but not required to assist them. The weight of the hoe and the length of its handle were always proportioned to the age of the child.

Early in the morning, except when the weather would not permit, the driver, standing in his door at the head of the street, would awaken the field hands by blowing a horn, though on some plantations a bell was rung. They were awakened early enough to give them time to cook their breakfasts and to put up something for the midday meal. Then they all gathered at some central spot and started for the fields, the driver leading the way. As they went along in a gang there was usually much talking and a good deal of jesting. Some were very voluble, but in reality said nothing, scarcely understanding themselves, while others were quite full of humor and said many witty things. The men rarely joked the women, but the women had much to say to the men, seeming to make fun of their looks, while a few were sullen and morose and had little to say. These had no "mornin' " for anybody.

In the spring and summer, the men and sometimes the women went to the field without shoes, the latter usually wearing cotton leggings fitting closely and coming nearly to their knees. Their dresses came just below their knees, and were tied just below the waist with a cord, thus forming a roll. When the dews were heavy, the men rolled up their pants to keep them dry, and some would wear their caps while others left them at home. The women, with few exceptions, wore colored handkerchiefs on their heads.

When the squares where they were to work were reached the line would close up, and all the Negroes would gather in a crowd. Many of them promptly sat down on check-banks, all waiting for their names to be called and to be "set" their tasks, the field having been laid off in half-acre tasks. "Fall en yeh, Lizabet." "You, Scipio, tek dat tas' hed of Tom." "Gal, ona fall en behine Isrul." "Chillun, ona tek de tas' longside ob yo ma, and mine don't fool wid me today." Such orders as these have been heard thousands of times on rice plantations, just as the sun, coming through a bank of clouds in the east, began to dispel the heavy white fog which often lies close to the ground.

The field hands were classed as half hands, or full hands, and the work of each, in both character and amount, was what long experience had shown could be done, and done thoroughly, for in slavery days great stress was laid on the quality of the work.

It was always the custom during the spring and summer months to allow the field hands to finish their work early enough to give them at least two hours by the sun in order that they might work for themselves, or if they did not care to work, they were allowed to do as they pleased. During the winter months opportunity was given them to gather firewood and to grind corn. To all full field hands, who would make use of it, high land was allotted and they were encouraged to plant crops for themselves, and this many of them did. In fact, on rice plantations probably more Negroes planted their own crops on a small scale when they were slaves than they did after they became free. Negroes who did plant for themselves after freedom usually wanted to plant more land than they could cultivate, thinking that by so doing they would be more independent and could work or not, as they chose. The result was that many of them were continually in debt and at the end of the year had nothing.

Each year during November or December woolen cloth was distributed to the Negroes, and, in May or June, cloth of a lighter material. This cloth was of Australian wool and was of excellent quality. The color of most of the cloth was gray, though some was blue. All of it was imported from England. Every field hand was given five and a half yards of gray cloth, and a smaller quantity for each of his or her children. For a baby, a mother received one and a half yards. The drivers, carpenters, and other head men, to distinguish them from the rest, were allotted six and a half yards of blue cloth and one of white. They were also given overcoats and felt hats. The men among the field hands received caps, and the women plaid handkerchiefs known as bandannas, which they tied around their heads and wore constantly. Woolen blankets were given to the slaves when needed.

In the fall each Negro was provided with a pair of shoes of substantial quality. There was no haphazard distribution of these shoes; each pair was ordered to fit a certain individual. To accomplish this, a small cedar stick, neatly made by a plantation carpenter, was given to every slave. After measuring the length of his or her feet, the slave would cut the stick accordingly, and notch it to indicate the width of the feet. These sticks were sent to the factor in Charleston, along with the order for the shoes.

The slaves set great store by their shoes and usually took the best care of them. I have heard that once on a very cold morning, when the ground was frozen, a Negro was seen walking along the road barefooted and carrying a new pair of shoes in his hand. Asked why he did not wear his shoes, he replied, "Well, oonuh see dese duh me shoesh; me feet dem blonxs tuh Maussuh."

Regularly one day each week the slaves were rationed. They were given corn, sweet potatoes, rice, and syrup. Each adult received four quarts of corn, and one-half of this amount for each of his children. When sweet potatoes were given plentifully, the amount of corn was slightly reduced. Until the time of Charles Heyward, they were not given meat; this they were expected to provide for themselves. In order to do so, they were allowed to raise their own hogs and were given the privilege of hunting and fishing. Fish were very plentiful in the Combahee, and often a narrow mouthed black bass could be caught which would furnish as much food as the average Negro family would require in one day. In addition to this, most of the slaves raised poultry and often sold eggs.

One would naturally suppose that since the slaves received no wages, they did not have any money, but this was not the case. They always managed in one way or another to have something to sell, and to do some trading. Adjacent to every two or three plantations there was a crossroads store, with which the owners of the plantations had nothing to do, but where the Negroes could trade. At these stores they purchased their tobacco and such small articles as they might need or fancy. Although it was the rule that they could not leave the plantation without permission, this was rarely refused, especially to go to the store, and often one Negro would do the buying for several. It was at these little near-by stores, during the days of slavery, that rice-field Negroes formed the habit of buying groceries in very small quantities, and from this habit they never recovered. For many years after the Civil War, though they had enough money in their pockets to buy at one time a week's supply of provisions, they much preferred to buy what they wished each night, by the nickel's or dime's worth. "Gimme tree cent wut tobacco; lemme hab five cent sugar, want ten cent wut butt meat," was the way they generally dealt in the stores, and this they do to a large extent even today. In making their purchases they were quite good at addition, keeping the amount of a number of small items in their heads. They could seldom understand the principle of debit and credit, though to one kind of "scredit," as they called it, they have always been particularly partial.

If, for instance, they had an account with the plantation store, and a portion of their wages was withheld on Saturdays and placed to their credit, some were unable to understand, if they purchased a pair of shoes for $1.50 and later a similar amount was withheld and credited on their account, why the item of the shoes should still appear on the account. "Enty de dollar and a half wipe out de shoes? Den you should scratch um off," they would say.

On every plantation before the Civil War, there was a grist mill for general use, but most of the grinding of corn was done by small grinding mills, which were portable and which the Negroes moved from house to house, one mill being allowed to every five families. These grinding mills consisted of a flat round stone, which was placed on a stand about three feet from the floor. Another stone was laid on top of it and was worked on a pivot by a crank which hung from the ceiling. The crank was a long wooden pole, one end of which was fastened to the ceiling directly over the center of the stone and the other end to the edge of the upper stone, the turning being done by hand. From the corn placed between these stones both grits and flour were made, the flour consisting of the eye of the corn, and the grits of the remainder of the grain.

Not until after emancipation were the Negroes on Combahee plantations given an acre of rice land to plant for themselves. When their crop was harvested, the streets in the settlements were busy places for at least a week. Throughout the entire length of the street Negroes could be seen with flail-sticks, beating the rice off the straw, while with mortars and pestles others pounded the hulls off the grain, and the children with fanner baskets carried the rice into their houses to be safely stored. When working for themselves, the Negroes never seemed to tire. The flail-sticks fell fast and vigorously, and the pestles never missed a stroke. It was their rice and they wanted to get it under shelter as soon as they could. They trusted neither the weather nor each other, for rice was their favorite food.

I have often ridden through settlements on my plantations when the Negroes were threshing their own rice and wished they would work as hard for me as they did for themselves. But this they never did, even when they were given "tasks."

Sunday was always a day of rest, and frequently church services were held on the plantation by white ministers, whose salaries were supplemented and traveling expenses paid by the planters. There were then not as many "local preachers" among the slaves as sprang up after they were freed; but the preachers they had were allowed to hold prayer-meetings at night, though at these meetings there was not as much loud shouting and promiscuous praying as there was after the Civil War, when many of the men on the plantation became either "class leaders" or "locus pastuhs."

During the days of slavery a considerable number of slaves on the rice plantations of South Carolina were members of the Episcopal Church, as most of the ministers who held services for them belonged to that denomination. When the slaves were declared free and were no longer preached to by white clergymen, they very rapidly established their own churches of other denominations, so that today in South Carolina few Negroes are members of the Episcopal Church.

The rule that the slaves should not leave the plantation without permission, I am sure, was often broken, especially at night. When they were back on time for work in the morning, the drivers said nothing. The Negroes were never kept under guard, and it was not difficult for them to obtain permission occasionally to leave the plantation during the day when it did not interfere with their work. When night came, they were expected to be in their own houses, but no doubt this rule was often violated, by the men especially. The enforcement was left largely to the head drivers, though the overseer would try to see that it was carried out. Usually at night the drivers were in their own houses, and a slave who happened to be interested in their whereabouts knew they were there and realized if he could "ketch back fore day clean" the risk was small, for there were so many of them and all were dressed alike. The county patrols had little or nothing to do, for seldom did a Negro on the Combahee try to run away.

21

THE HOUSE SERVANTS

EVERY former slaveholder in the South has now been called upon to give an account of his stewardship, and nearly all of the Negroes, too, who were slaves, have passed away. Both master and man must stand before the same judgment seat, and in most cases when they meet there they will be glad to see one another, for the bonds which bound them together on earth, strange as it may seem, were strong.

The house servants of the old South lived in daily contact with the members of their owner's family, and they considered themselves almost members of the household. Whenever any distinction came to the family through politics or success in any line, the slaves were proud of it, bragged about it, and gave full play to their imagination. This was especially true of the house servants. These Negroes never thought of working in the fields. They considered themselves in a class entirely apart from the field hands. No worse threat could be made to them than that they would be sent to work in the fields.

The children who lived in the Low Country of South Carolina after the close of the Civil War, and whose families could afford to employ some of their former slaves, will never forget the old Negroes. They were their confidants and advisers, their allies in time of trouble, and often the children of these servants were their favorite playmates. Some children used to love their nurses, those faithful old black maumas, as much as they loved their parents. These maumas were usually stout women, most of them not being given to much exercise. They were always neatly and cleanly dressed and wore colored bandannas wound around their heads. I once took one of them with my family on a trip to Virginia and she insisted upon wearing a red and white bandanna under a black straw hat.

In Charleston, and other cities in the South, these old maumas much preferred talking to walking. They delighted to sit and talk, three or four of them together, for hours at a time in some shady place in summer and in a sunny spot in the winter, while their charges played within easy hearing distance. The children obeyed them too, for they soon were trained so that they did not have to be followed foot to foot. One great merit the nurses had was that they did not want every other afternoon and every Sunday off. Many modern inventions are in use these days to lighten a mother's work, but when it comes to minding children, nothing has been found to take the place of the Negro maumas of a generation ago.

Taking them as a class, the house servants of ante-bellum days in the South were much more intelligent than the generality of the slaves. They were carefully selected and also had far greater opportunities than the other Negroes. Their close association with white people likewise gave them a great advantage over the others. Nor did they speak Gullah to anything like the same extent as did the other Negroes. A butler in a Low Country family, though born on a rice plantation, used the Gullah about as little as do the Negro waiters in a Charleston hotel today.

When the slaves were emancipated, it was very hard for an old slaveholder to realize that his Negroes, and especially his house servants, no longer belonged to him, and the Negroes, I am quite sure, felt the same way. Their former owner's home was the only one they had ever known, and they thought they belonged there. When they had to find employment elsewhere, if in the same city, they usually continued to stay on in the old servants' house in their former master's yard. The feeling that she still owned her former slaves is very strongly and yet amusingly expressed in the will of an old lady of Charleston, made nearly ten years after her slaves had been declared free. This will is now on record in the Court of Probate in that city. In it she says: "The following Negro slaves [naming them] who are still my property, but whom it has been sought to take from me by a high-handed and confiscatory government, I hereby devise and bequeath, as follows: . . ."

The Negroes whose names she mentioned were evidently house servants who had either remained in her service or had continued to live on in her yard after their freedom. I fear her bequest was not very valuable.

The house servants in slavery times had a very easy time of it. Many of the older men died from dropsy. An old Negro, who had been a butler before the war, was once asked if he had had to work very hard.

"Not so berry," he said. "De wust job I hab was walkin'

to answer de front door bell. It just seem to keep a-ringin' all de time. It sure kept me a-movin'."

One reason the servants of ante-bellum days did not have to work hard was that in the homes of most of the wealthier planters, there were so many of them. I do not see how the mistress of the house ever kept up with them all. There were always, of course, a cook, a butler, a coachman, a nurse, a chambermaid, a seamstress, and a laundress, and in some households each of these had one or more assistants. The butler, for instance, in some families, was not required to wait on the table; his was the responsibility only. Two young butlers did the actual waiting. The coachman never touched the horses; he only handled the reins. A hostler or two hostlers curried and attended to the horses. Of course the reason for this latter was that when the coachman, seated on his high seat in his livery and beaver, drove the family carriage on a summer's afternoon around the Battery in Charleston, the salt breezes from the sea should convey no suggestion of the stable.

Having a number of house servants was the greatest luxury the planters of those days indulged in, but they did not look upon it as an extravagance. They owned the Negroes— why not have them as house servants? There was not much outgo visible in feeding and clothing them, but that outgo must have been there just the same. It was like the indirect taxes which we pay today. In other respects, the planters lived rather simply and unostentatiously, but to be waited on was with them, as it is with us now, a very pleasant luxury, and luxurious habits are readily formed and easily become fixed. A very respectable Negro of Richland County told me that before the war he belonged to a wealthy old bachelor, a planter, to whose wants he regularly attended. He said he had to hand him his toothbrush every morning.

A few months ago, I saw again a letter which was written to me by my father when I was four years old. (It was written from Charleston and I was in Greenville at the time with my mother.) As I read it through, I wondered how many fathers in the South these days would write their children such a letter, filled as it was with the doings of their Negro servants, and especially their servants' children. My father did this because he knew I would want to know all about them.

In this letter my father, after mentioning that the goat and the ice cream man were missing me, said that he had covered up my rocking horse, so Johnnie and Susan, children of old Paul, the butler, could not ride it. (They used to ride it with me all the time when I was at home.) He also told me about what the other Negroes—those still living in the yard but working elsewhere—were doing. He then promised to get an alligator from Combahee and have it tied securely on the piazza so I could shoot at it with my toy pistol when I got back. Even then, he must have foreseen that I was to follow in his footsteps, plant rice and live among the Negroes and the alligators. My father closes his letter by saying, "Paul is ringing the dinner bell and I must go down stairs."

When I read about "Paul ringing the dinner bell," my childhood days seemed to come trooping back to me, through the mist of years. I can see old Paul just as I saw him when I was a child, when we lived on Meeting Street, near the Battery, in Charleston.

Paul was a large, portly Negro, with a round head, and very black, but unlike Sam Johnson, the founder of Calvary Baptist Church in Columbia, there was no Madagascan blood in him. As I think of old Paul now, I know that he must have been a pretty hard case, but to me in those days he was attractive, and I was his pet. The greatest virtue Paul had was his unswerving loyalty to my father, whom he idolized; and his worst fault was that he was very fond of whiskey. He was older than my father, and before the war, when my father went hunting, Paul always went with him to carry the game, and after the war he remained as butler until my father's death.

Though I cannot recall that part of it, I know now that Paul must have been a thoroughly trained butler, for he could perform his duties when drunk as well as when sober. In those days, it was the custom in Charleston to have late dinners, about four o'clock, and very light suppers, the latter being handed around in the drawing-room on a tray. Paul had a very large silver tray with a leather strap to go across his shoulders. When he had been drinking and seemed unsteady in the pantry, his wife, Betsy, and his children, would harness him to the tray and assist him to the drawing-room door, steady him, and then start him through the door. Once they got Paul into the drawing-room they had no fear; he was certain to make the rounds, not only successfully but with much dignity. They would wait for him outside the door, help him into the pantry, and unharness him from the tray.

It amused my father greatly to tell stories about Paul in his presence, especially when he saw he had been imbibing a little too freely. His favorite story, told more to hear what

Paul would say than for its own sake, was that one day he was shooting snipe, with Paul along as usual, and that during the morning Paul picked up a fish in a ditch and put it in the game bag. When they stopped for lunch, my father said he told Paul to count the snipe. He did so, and announced:

"We hab t'irteen snipe countin' de fish."

Old Paul would look very glum, but say nothing. Soon he could be heard mumbling to himself very audibly: "I never tell Mass Barnwell no sich t'ing. I tell um we hab t'irteen snipe 'scusin' de fish."

Poor old Paul—he died from whiskey and dropsy. His wife, Betsy, a bright intelligent little woman, lived on in Charleston until a great age. As far as I know, all their children—my black playmates of those days—are also dead. Betsy outlived them all. Some years ago, when I wanted to know whether or not a portrait of my mother was a good likeness, I sent for Betsy.

22

TWO FORMER SLAVES

SEVERAL of the slaves of Nathaniel Heyward, after their freedom, reflected great credit on their race, but two especially I wish to mention, because both are outstanding examples of what a Negro, born in slavery, was able to accomplish. One became a college president, and the other, besides being a devout Christian, became a forceful preacher, preaching to the white congregations in the North and to his own race in the capital city of his state, and worked with such zeal that his name is still remembered.

Some years ago I was in my office in Columbia when I was told that a Negro man wished to see me. When he came in I was much struck with his appearance. He was a fine looking old Negro, tall and dignified in his bearing, dressed in a Prince Albert coat and carrying a beaver hat in his hand. I wondered who he was and on what business he had come to see me. He told me his name and I asked him to be seated.

We talked generalities for a while, and then he proceeded to tell me that he was the president of a small Negro college in South Carolina. After a little while I asked him what I could do for him, and was certainly surprised when he replied, "I have come, Sir, to get you to tell me how old I am."

"Why," I answered, "how in the world should I know? I have never seen you before."

"Yes," he said, "that is true, but haven't you got 'the book'?"

And instantly I knew that he must have been a former slave of Charles Heyward, for I remembered how often similar requests had been made of me during my rice-planting days, and strange to say, I happened that very day to have in the office Charles Heyward's records of his "emancipated slaves." Getting out the old record, I said to him, "To which plantation did you belong?"

"To Rose Hill," he replied, and added, "I am the son of Clarissa, the child nurse."

I soon found his name and told him his age, and also that I had often been told what a fine woman and nurse his mother was, which seemed to please him greatly, and then we had quite a long talk. He told me he had left Combahee shortly after he was freed, and inquired about some of the Negroes whom he remembered, wanting to know what had become of them, whether they were alive or dead.

I have sometimes wondered whether, if anyone else had been present, that college president would have asked me the question he did. I doubt if he would have. He did not mind asking me, for he knew that we would understand each other.

Several years ago, on a late winter's afternoon, I was driving through Richland Street in Columbia with a cousin of mine who for years had lived in another state, and who was then past his three-score years and ten. As we came to Calvary Baptist Church, one of the largest Negro churches in the state, my cousin asked me to stop and go with him into the yard in front of the church, where I could see two tombstones standing side by side. One headstone, I found, was to the memory of the Rev. Samuel H. Johnson, the founder and first pastor of the church, and the other marked the grave of Kate Johnson, his wife.

For several minutes my kinsman stood silently before these graves. His expression, I noticed, seemed to soften, and I could see that his thoughts were in the past. Memories of long ago came back to him as he stood and looked at the two headstones in the deepening twilight.

Seeing us standing there, the pastor of the church came out and asked if he could be of any service.

"Did you happen to know anything about the Rev. Samuel Johnson?" he asked my cousin.

"Did I know Samuel Johnson!" my cousin exclaimed. "Why, he is one of my earliest recollections. When I was a boy, Sam and I were the best of friends. I taught him to read and write, and his wife, Kate, who is buried there, was my nurse."

The preacher was surprised and interested and he asked my cousin to come to his church that night and talk to his congregation about the founder of their church. Few if any of his people, he said, had known Samuel Johnson, but the reputation he had left was such that the congregation greatly revered his memory.

My cousin thanked him, but declined, and as we rode home he told me the story of Sam Johnson, who was born a slave and died a useful citizen highly respected by both whites and blacks, and one whose work among his own race can be credited with the saving of many souls. The inside story of Sam

Johnson's life has never been told and is probably known only to my cousin, who has now passed away, and to me.

Samuel Johnson was born in 1813 on either Blanford plantation—one of the plantations owned by Nathaniel Heyward on the Combahee—or in Charleston. His name appears in the appraisal of the slaves made by the executors of the estate. Blanford was one of the plantations left in Mr. Heyward's will to his grandson William Henry Heyward, the father of my cousin James B. Heyward, who was Sam Johnson's boyhood playmate and who told me the story of Sam's life.

Many years of Sam Johnson's life were spent on a rice plantation in winter, and in Charleston during the summer, for Sam was the family butler. He was a Negro of Madagascan blood, with a good mind, and very straightforward. He stood about five feet ten inches, was inclined to be somewhat stout, had a head of thick bushy hair, and was of a gingerbread complexion. He spoke with considerable fluency, and in his later years, when he preached to large congregations at home or elsewhere, he commanded their close attention.

Slaves from the island of Madagascar were comparatively rare in South Carolina. I doubt if any were imported into Charleston direct. Those owned there were probably purchased from the colonies farther north. The Madagascan Negroes commanded much higher prices than the other slaves, for they were considered more intelligent and dependable. They were also lighter in color and did not use, to any great extent, the Gullah language. Altogether, they were of a higher type and hence were much sought after as house servants. When these Negroes were properly trained, they were probably the most efficient servants to be found.

And Sam Johnson was one who was well trained. When a boy he was the assistant to the butler, and when quite a young man he was made butler for his owner's household. This place he kept for nearly thirty years, and during all that time had the full confidence of his master and mistress, and the affection of the children of the household. There was nothing these children would not do for Sam, and he could keep in close touch with them, for he married Kate, their nurse, and in those days children were left largely to the care of their Negro nurses.

Now Sam had one great ambition, and that was to learn to read and write, so that he could study the Bible and teach it. Shocking as it may now seem, in those days the laws of South Carolina prohibited the teaching of Negro slaves, and it looked as if Sam had little chance to attain his ambition. Fortunately for him, however, there was a friend at hand, one of his master's sons who was then quite a lad.

It seems that my cousin's mother very properly believed that children should not drink tea or coffee, both of which the boy wanted. Sam had both, and Sam wanted to learn to read and write. So they plotted together, formed a plan, and made a deal. Sam was to give his own tea and coffee to the boy, who in turn was to give him lessons in reading and writing. Such an arrangement was easy to carry out, for Sam's wife, of course, would not tell. They would stealthily meet in some safe place and together would study my cousin's lessons for the next day. In a short while Sam was able to read the Bible, much to his delight.

After a while it was discovered that Sam could read, for he was often seen studying his Bible. No objection was raised nor was it ever suspected how his learning had been acquired. Shortly after this he joined the Baptist church, and later was ordained a preacher by the Reverend Richard Fuller, a Baptist minister who preached to white and Negro congregations in and around Beaufort. This was all before the Civil War.

My cousin's father owned a number of slaves, and these, early in the war, he moved to the neighborhood of Columbia, and rented for his family a residence in the city; it was thus that Sam Johnson happened to come to Columbia.

All during the war he continued to act as butler for the family, though his duties as such must have been very light, for there was then very little food to be had and the meals Sam served were meagre. He had ample time to teach a class of Negroes and often at night he preached to a number of his own race. All the while he was loyal to his owner's family, and the women and children felt that in him they had a faithful protector.

Such he fully proved himself to be before the war was over, for when General Sherman captured Columbia, and his troops all but burned the entire city, Sam Johnson went among the drunken soldiers and told them that the house where his owner's family lived was his house, and begged them not to destroy his home.

It was thus due to Sam, my cousin said, that the house his father rented is standing today; and scarcely two blocks away is Calvary Baptist Church, built shortly after the war, through Sam Johnson's efforts.

When slaves in the middle section of South Carolina were declared free, those from rice plantations, with few exceptions, returned to their former homes. Sam Johnson, however, was an exception. He decided to remain in Columbia, where during the war he had formed many attachments. Entering actively on his work as a preacher, he made such progress that he decided to build a church of his own. Finding that he could not obtain sufficient money in sorely stricken Columbia to erect the kind of building he wanted, he managed to go to Baltimore, where he was invited to preach to white congregations. There he pleaded so earnestly that most of the money required was given him, and he came back and built the church which today stands as his monument.

My cousin ended his story by telling me that while Sam Johnson lived, he never came to Columbia without going to see him at his home near Calvary Church. In a little sitting-room they would talk over the old days and the changes the years had brought. With them the ties of the past still held fast, as I can myself well understand. Not even death and the long flight of years could break them. My kinsman's face proved that, as he stood before the graves and read the wording on their tombstones.

23

SQUIRE JONES AND MR. JAYCOCKS

A NUMBER of years ago I dined at the home of a friend in New York City, and, among several guests, with one exception I was the only Southerner present. While at table, during a lull in the conversation a lady remarked loud enough to be heard by all, "I have always understood that Southern planters were cruel to their slaves."

The lady did not know that I was a South Carolina rice planter and that my forebears had been slaveholders, and hence she was quite embarrassed when our host very pleasantly informed her of that fact, and she sought to modify her statement. "It was not, of course, so much the planters themselves," she said, "who were cruel to the slaves, but their overseers. Certainly it was so represented in *Uncle Tom's Cabin.*"

I told her that I would not like to decide the question she had raised, for my present overseer was born in the City of New York, and when I returned home I would certainly tell him what one of his countrywomen had said about men of his calling, and I knew he would be greatly amused. I did not, however, mention the fact that Mr. Jaycocks had been taken to Charleston when he was three years old and at sixteen had served for two months in the Confederate Army. And I then proceeded to tell my own experience with two overseers, the only two I had come in close contact with.

Mr. Jones, as I have stated, came as overseer to Amsterdam and Lewisburg plantations January 1, 1855, shortly after the devastating storm of 1854, and remained with my grandfather, Charles Heyward, until the latter's death in 1866. Of those eleven years, he spent three at Goodwill managing the Combahee Negroes whom he had taken there early in the Civil War. Shortly after the slaves were freed he retured to his home in the pineland near Combahee, where he awaited in January, 1867, the coming of my father, in the meantime keeping him posted as to conditions.

Upon my father's arrival, Mr. Jones promptly joined him on his plantations and remained there until January, 1889, when he retired on account of old age, being succeeded by his son-in-law, William Jaycocks. The length of service of the two combined totaled sixty-one years, a most remarkable record.

When I began rice planting Mr. Jones was still the overseer, and many talks he and I used to have, as we rode together through the fields, or sat on a winter's night before a log fire in the dining room of the little cottage at Lewisburg. Many of the things he told me I well remember, especially what he said regarding the treatment of the slaves on the plantations of Charles Heyward, whose management of his slaves followed closely the methods pursued by his father, Nathaniel. What Mr. Jones said was often confirmed by Negroes who had belonged to both Charles and Nathaniel.

As to the treatment of their slaves the Heyward family, according to Mr. Jones, never enjoyed the reputation of having been especially humane, nor was it ever charged that they were in any degree cruel. Their management of slaves was fairly typical of the way other South Carolina rice planters treated theirs.

There were no doubt some exceptions among the slave-holders, but these were rare, and in nearly every instance were due to absentee landlordism. Let one read the accounts of cruelties practiced on a few Southern plantations years ago, and it will be seen that, where the charges were in any degree substantiated by facts, this was the underlying cause. In South Carolina those who ill-treated their slaves were generally looked down upon.

That slaves were occasionally punished, Mr. Jones readily admitted. Very little punishment, he claimed, however, was necessary, because the Negroes had lived with the same family for generations and a kindly feeling existed between their owners and themselves. If a man or a woman was negligent in his or her work, in all cases punishment was administered by the driver. When the driver thought it deserved he inflicted it in the field with a leather strap fastened to a wooden handle, and applied to the slave's bare shoulders. It was never the business of an overseer on a rice plantation to punish a slave.

I have, in the past, discussed this subject also with men, both planters and overseers, who were familiar with rice plantations before the Civil War, and practically all of them were in full agreement with Mr. Jones. I recall one instance especially when I was told by a trustworthy gentleman who had lived on a South Carolina rice plantation for six years before the Civil War, that he was sure that he had never seen a driver use his strap on a Negro more than a dozen times, although there were seven hundred slaves on the plantation.

To many today the punishment I have described may seem somewhat shocking, but those who some fifty years ago attended private schools in either the South or the North would not so regard it. I went to two schools in Charleston where the pupils would have gladly seen the rattans which were used on their backs and legs by the teacher exchanged for the leather straps once used on the slaves.

I knew only one of my great-grandfather's drivers, one who, the older Negroes on the plantation said, was freer with his use of the strap than any of the other drivers. This was a small Negro named Jack, although the Negroes before the War nicknamed him "Wasp." He lived to be an old man and used to come every now and then to see me, and would tell me that the Negroes I was working "ent wut killen." He further used to express the wish that, old as he was, I would put him in charge of them for a short while, and this I sometimes threatened the Negroes to do.

"Great Lawd, Maussuh," the older ones would say, "no nyuse fuh sen' ole 'Wasp' fuh foreman ober we on dis plantashun; 'e too ole—'e can't jump cross ditch."

I would tell them that I could easily remedy that. I would have old Wasp carried across the ditches.

"Now, Maussuh, do lef dat ole nigger weh 'e day; 'e bad tummuch," they would laughingly reply.

Yet, I never noticed that old Wasp seemed at all unpopular when he mingled with them in the commissary. Even the old Negroes seemed to have forgiven him.

There was no kinder-hearted man than Mr. Jones, "de ole Boss" or "de ole Squire" as the Negroes called him. He would get mad with them and sometimes cursed them, but there was never any anger in his voice. Often when he became angry with a Negro, I have seen the Negro walking away and laughing to himself, saying, "Ole Boss lub fur cuss me, but 'e ent mean um. 'E pay fur cuss me." Lucius, the engineer of the threshing mill, who once thought himself very efficient, frequently came into conflict with "de ole Boss." Whenever anything went wrong with the machinery, a bell was rung as a signal for Lucius to stop the engine. When awake, this he usually did, but he would then immediately pick up his monkey-wrench and begin recklessly to take off nuts and bolts, regardless of where they were. Mr. Jones would approach him from behind, and, laying his hand on his shoulder, would say, at first in a very soft voice, "Lucius, my dear fellow, stop. Stop, Lucius!" Then in a louder voice, "Lucius, you damned old fool, what in hell are you trying to do?"

The Negroes on Amsterdam and Lewisburg plantations were very fond of Mr. Jones, for he had managed them for years and was always kind to them. But he always liked a Negro in what he called "his place." He never liked to see him putting on what he considered "airs."

The first year I planted rice, one spring morning as I rode through the fields, I met Mr. Jones riding a check-bank to meet me. The old man looked excited; his face was white and I feared something serious was the matter with him. As soon as he reined his horse he said, "I have been on this plantation thirty-odd years but I can't stay here another day. I'm through."

I asked him what in the world had happened.

"Happened, hell," he said. "Saby Small rode to the field this morning on a bicycle, to hoe rice, and Dove Campbell asked me to ask you to get her some vanilla! She wanted to make ice cream. Now, if these Negroes are going to ride to work on bicycles, and eat ice cream, my day is over."

I wonder what the old man would have said had he lived long enough to see Negroes he had known from their childhood driving around on Combahee in their automobiles.

After living with these Negroes for thirty-three years, when Mr. Jones resigned they hated to see him go, and the parting hurt him as much as it did them. As long as he lived, he would drive frequently to Combahee and his interest in the plantation and the Negroes never ceased.

The day the old Squire passed away, there was a deep stillness at Amsterdam and Lewisburg; all work ceased and the Negroes prepared to attend his funeral. He was buried in a little graveyard in the pineland, where many of his kinsfolk slept, and the old Negroes, former slaves, standing by his grave, knew that they had lost a lifelong friend.

William Jaycocks, who was born in the city of New York, was brought by his mother to Charleston when he was a child. His mother kept house for my father after the death of his wife, and young Jaycocks grew up in my father's family at Goodwill plantation.

When the Negroes of Charles Heyward were brought to Goodwill, young Jaycocks soon knew them all, and never did he forget one. When my father went to Combahee after the close of the war, Mr. Jaycocks went with him and remained with him until his death, helping him on the plantation. He then became manager on a rice plantation on the Savannah River, where he remained seventeen years, until he returned to the Combahee where he and I, for twenty-five years, fought it out until the bitter end.

For that length of time, no two men's business and personal relations could have been closer than ours. Together we experienced good years and bad, good harvests and West Indian hurricanes. Some years we could see some future for rice planting in South Carolina, and then again we could see none. Often we were forced to see "the handwriting on the wall."

Mr. Jaycocks was one of the very best managers of Low Country Negroes I have ever known. Intuitively Negroes trusted him; they knew that he was fair. When he "docked" them, as they called it, for bad work, scarcely ever did they complain. He never swore at them, even pleasantly, as did Squire Jones, and seldom lost his temper. When it came to stopping a break in the river bank, if Mr. Jaycocks could not stop it, no one could. He ought to have been good at stopping breaks, for during the years we worked together he certainly had ample experience.

When I think of Mr. Jaycocks, as I often do, I picture him to myself on horseback, as I daily saw him. He was a splendid horseman. I never knew a horse or mule to get him out of his saddle, nor to run away with him in a buggy, and in those days, it seems to me that when he and I were not stopping breaks we were on horseback or driving in a buggy or roadcart.

Often during the harvest season, before the cause of malaria was traced to the mosquito, Mr. Jaycocks and I would drive at night to his summer home in the pineland, and return to the plantation early the next morning. I shall always remember those drives, of ten miles each way, especially when the nights were dark, and I am sure that in the Low Country of South Carolina the nights are blacker than anywhere on earth, certainly under the live oaks.

I used to accuse Mr. Jaycocks of being able to see better on a dark night than he could in the daytime, and he would accuse me of not being able to see at all. Once he certainly proved it. It was one of the blackest nights I ever recall. We were driving a nervous, highstrung horse, and when we came to a gate Mr. Jaycocks gave me the choice of either getting out and opening it, or driving through. I decided that I would do the latter. When he opened the gate, he asked if I could see where I was driving, and I had to confess that I could not see the horse, much less the gate.

After a day's work, I would find those long drives in a buggy very tiresome. I could find some consolation, however, when I thought of an old Negro on Combahee, who under similar circumstances proved himself somewhat of a philosopher. He waited to be paid off one afternoon during the harvest season, and when he started for his home in the pineland, in his little ox-cart, with his grandson along, it was night and so dark he could scarcely see. The ox, weak from standing in the hot sun all day, swayed from side to side against the shafts. To the old Negro, hours seemed to pass. Occasionally he dozed, then awoke with a start and said a few words of encouragement to the ox. Once, suddenly waking and not being able to see where he was, he felt uncertain whether he was moving or not, and decided to make an investigation. He shook and woke up his grandson.

"Boy," he said, "jump out. Is you out? Well, den put you han' on de wheel."

"I'se got um," said the boy.

"Berry well den. De wheel duh turn?" The boy hesitated a moment.

"Yes, Grandpa, 'e duh turn."

"Jump in, boy," said the old darky, much pleased, "we is gwine home."

Mr. Jaycocks certainly knew Negro nature. For fifty-odd years he was closely associated with the Gullah Negroes of our coastal section. He not only seemed to know what a Negro was thinking, but could himself think like one. This perhaps accounts for the Negroes' understanding him and appreciating the fact that he was interested in them. No Negro on the plantation was ill but Mr. Jaycocks went to see him and did all he could for him. If a doctor was needed, he saw to it that one came. It was these little attentions on his part which endeared him so to the Combahee Negroes.

Close association with both Mr. Jones and Mr. Jaycocks, their experience with Negroes during slavery and after emancipation, the regard they had for the Negroes and the respect the Negroes had for them, fully convinced me that the slaves on our rice plantations had not been ill-treated or unduly punished. I might even go further and say that the rice planters of South Carolina and Georgia were, in most instances, not only fond of many of their slaves but even refrained from hurting their feelings by speaking of them as *slaves.* Knowing that the word slave, the world over and from the earliest times, had been looked upon as a term of reproach, most slaveholders, it may surprise many to know, never used that term, but spoke of them as their Negroes.

Old Negroes in my day never liked to be referred to as having been slaves any more than a self-respecting Negro of today likes to be called a "Nigger." Nor did the Negroes on the Combahee speak of their having been the slaves of their former owners. They always said that they had "belonged" to them, and let it go at that.

24

MY FATHER'S PREDICTION

MANY times have I thought of a prediction my father made in regard to the future of the rice industry in South Carolina and Georgia. This prediction proved to be true, although it was not entirely due to the cause upon which he based it. I have wished, too, that I had heard of this prediction many years before I did, for had I done so it might have changed the whole course of my life.

This prediction of my father's was made during the summer of 1868, while driving one afternoon from his plantation to the pineland with a cousin of his, a young man, who was just beginning his life as a rice planter. He told this cousin of his that he believed rice planting in our coastal section would continue to be fairly remunerative only as long as the generation of Negroes who had been slaves were able to work, and this, he estimated, would be about twenty years. He advised him at the end of that time to sell his plantation and give up the business.

I did not hear of this prediction until thirty-seven years after my father made it, and, strange to say, the very year when, according to my father, I should have sold my plantation, I planted my first rice crop. Some years later, I experienced the fulfillment of his prophecy.

It is generally thought that the decline of rice planting in South Carolina and Georgia began as an immediate result of the emancipation of the slaves, but this was not true by any means. Emancipation caused a loss to the planters of the entire capital invested in slaves, and in addition was directly responsible for a great depreciation in the value of their plantations. Notwithstanding this, however, the industry was continued with a fair degree of success for two decades after the close of the Civil War. A great many of the Negroes, after their freedom, remained on the plantations and in a short time adapted themselves to the new conditions, and many of them worked quite as efficiently as they had during slavery.

I shall first refer to why my father thought the culture of rice could not be carried on profitably for longer than twenty years. He believed that the second generation of Negroes, after slavery was abolished, could not be gotten to do their work properly, and to a large extent he was right. When the old Negroes who had been taught to work as slaves began to pass away, the work done in our fields deteriorated very greatly. No crop, and especially rice, can be a success unless it is thoroughly and properly cultivated, particularly when the work is done principally by hand labor. In our rice fields the hand hoe was always the most important implement, for, with the soil packed as the result of irrigation, it was essential during the season of dry growth to stir the ground thoroughly. Hence resort had to be made to the use of horse hoes.

During slavery, the hoeing of the crop by hand was so well and carefully done that scarcely a sprig of grass was left in the alleys between the rows. With the second generation after emancipation it became impossible to have a crop properly cultivated, and the same became true of harvesting. "Backache" became exceedingly prevalent among the Negroes; they could not stoop over to cut the rice with their rice hooks, and in consequence many of the heads were lost. In order to finish their tasks sooner, they began tying the rice in large sheaves, and poor stacking and much wet rice was the result. In many other ways the work done by the Negroes became more and more unsatisfactory, all of which seriously affected the yield per acre and also the quality of the grain. But what really caused the decline, and finally the abandonment of rice culture on our South Atlantic seaboard, was not only the failure of the next generation of Negroes after slavery to do their work properly, but also the competition and over-production of rice in the states of Louisiana, Texas, and Arkansas.

Beginning about the year 1885, the growing of rice in these states began to be felt by our rice planters in the marketing of their crops, and each succeeding year the competition became more serious. The consumption of rice throughout the United States being limited, these states soon produced more rice than the demand warranted, and for a number of years the price fell below the cost of production, though all their work in planting, cultivating, and harvesting the crop was done by the use of machinery.

Carolina rice, having long had a world-wide reputation, for some years sold at a higher price on the market on account of this reputation. Soon, however, wholesale merchants began branding rice from the southwest, and of an entirely different variety of seed, "Carolina Rice." As a result, the market for the genuine article became greatly curtailed, and our planters lost their one and only advantage.

Furthermore, the rice planters in the Southwest seldom suffered from tropical storms, nor were their lands subject to overflow by freshets, which sometimes occurred in our rivers. In South Carolina, the draining of fields and the destruction of forests in the upper part of the state caused freshets to become higher and more frequent. But on most of our rivers, storms did our rice far more damage than the freshets.

It was most unfortunate for the continuance of the rice industry in South Carolina and Georgia that during the last years of its existence the frequency of tropical storms should have greatly increased; and it was these storms which largely caused the last of our planters to abandon their plantations. Not only did the storms occur more often, but the damage they wrought was unprecedented in the history of the rice industry.

By the use of machinery, the planters in Louisiana, Texas, and Arkansas grew rice considerably cheaper than could the planters of South Carolina and Georgia. Except for the reputation of our Carolina rice, our one and only advantage was our lower cost of irrigation, but even this was to a considerable extent offset by our having to keep up extensive river banks, mend breaks in these banks when they occurred, and keep in repair our system of trunks. In much of the territory of the southwestern planters water was pumped up on the fields. This was done by large corporations, which charged a certain amount per acre, and the southwestern fields, being higher above the water level than ours, could be drained with more certainty, whereas we had to depend upon the fall of the tides which an easterly wind would prevent from falling to their normal depth.

Before the Civil War the growing of rice was begun on a small scale in Louisiana on the lower Mississippi River. It was planted on tidewater lands where conditions were very similar to ours. Not until a number of years later was its production undertaken, in the southwestern part of the state, on prairie lands which extended over a large territory. On these lands the planters had at first to depend to a large extent on rainfall to mature their crops, which made the planting of rice with them an uncertain business. Capital, however, was soon interested; large pumping plants were installed and canals and ditches dug; as a result, the growing of rice spread rapidly to similar lands in Texas and later into Arkansas. The corporations owning the pumping plants often owned large tracts of land also, and rented these lands to farmers from all parts of the United States. When a farmer failed to make good and had to give up planting, the corporation, through liberal advertising, induced another to take his place and thus the cultivation of rice was increased, with the result that the production exceeded the demand. Naturally, prices fell and the growers of rice in the South-west began to have their troubles.

When they made large crops, they were unable to market them at a price which would yield a profit, and this con-tinued to a greater or lesser degree until the World War, which greatly increased the demand for rice, not only in this country but throughout the world, and also greatly curtailed its export from Japan and other rice-growing countries.

The methods of planting, cultivating, and harvesting rice in the southwestern states differed greatly from ours and were considerably cheaper. Tractors there supplanted the mules we worked. It is true we used drills for planting, but our rows were only four inches wide, whereas there they were seven, which, with a narrow alley between, approached broadcasting the seed. Hence, the southwestern planters did not have to cultivate their crops two or three times as we did ours. They often sought to get rid of the grasses in their fields by planting them every alternate year, while we planted ours year after year, only occasionally resting certain squares, fighting our grasses with water and hand hoes. Also, by planting their lands every alternate year, they were better able than we were to get rid of "volunteer" rice.[1]

The rice planters in the Southwest used harvesting machines, which cut, gathered, and threshed their crops in the field and also sacked it ready for shipment. Buyers, representing pounding-mills located adjacent to the plantations, purchased the rice, which, when it was prepared for

[1] Rice known as "volunteer" rice has a red instead of a white inner cuticle, and for years the United States Department of Agriculture and the rice planters themselves in the Southwest thought it was a distinct species of rice, but in this they were mistaken. When a grain of rice is shattered from the head in being harvested, if it does not rot, isn't picked up by the birds, and gets under the soil, especially in clay lands, it sprouts the next year and matures with the other rice. When the next crop is cut, the heads from the volunteer grains shatter greatly when shaken by the reap-hook. Should any of these grains for a second time survive and grow when they mature, their inner cuticle will have become red. This process often continues for years. This not only lessens the yield but the volunteer grains get mixed with the crop, and, the red cuticle being much harder for the brushes in the pounding-mill to remove than the white, many of the latter are broken, lessening the value of the rice.

market, the mills undertook to sell. Their method of marketing rice differed greatly from that pursued by the rice planters of South Carolina and Georgia. We shipped our rice in vessels to the pounding-mills in Charleston, Georgetown, and Savannah, paid toll for having it pounded, and the mills stored it free of charge until it was sold.

For more than a hundred years our planters sold their rice through factors to "rice dealers," as they were known, and these in turn distributed it through wholesale merchants to the retail trade. The rice dealers had their representatives in the larger cities in the East and Middle West, and there was an agreement between them and the factors that the latter would not solicit the wholesale trade. Such an agreement tended to maintain prices, for under it the factors, controlling the rice, could to a certain extent uphold prices.

All of this was changed, greatly to the disadvantage of our rice planters. When rice from the Southwest, and also foreign rice on account of the lowering of the tariff, began to flood the markets of this country, rice dealers in Charleston and Savannah, whose principal business had been selling Carolina rice, were forced to compete with local brokers offering rice from Louisiana and Texas, as well as foreign rice.

Despite this change in their business, however, rice dealers in Charleston and Savannah still insisted that they should be the intermediaries between the factors selling Carolina rice and the wholesale merchant, and, when the markets in these cities became, as they did in some years, glutted with rice from the Southwest and with foreign rice, our planters had a very poor chance to dispose of their crops at a fair price. The pounding-mills in Texas were far removed from the rice-consuming sections of this country, and hence were not well posted regarding the markets, whether they were strong or weak, and as a result wholesale merchants often dictated prices. The mills had to dispose of the rice, and the brokers usually accepted the best price offered. After the year 1900 every firm of rice factors in Charleston and Savannah had been forced to give up business, and Carolina rice was sold only through a few brokers.

These men were efficient men, but they were greatly handicapped by the attitude of the rice dealers. I have never forgotten how discouraged I felt when one day my broker told me he had not been able to sell any of my rice. No dealer had called at his office for a month. He went on, however, to say that he was hopeful, for a friend had told him he had heard a gentleman remark that a certain rice dealer had been overheard to say he expected to be in the market for some Carolina rice within the next two weeks!

Returning to the plantation that afternoon, I conceived what I thought was a bright idea. Instead of our putting Carolina rice, as we had done for years, first in three-hundred-pound barrels and later in one-hundred-pound sacks, why not put it in small paper packages, with its merits advertised on each package, and sell only to the retail trade? Full of hope, I went to New York to see how my plan would take with the trade there.

Immediately upon my arrival I called at the headquarters of a large chain of retail grocery stores and asked to be shown to their rice department. The official in charge received me most pleasantly, and I asked him if he bought

Carolina rice. He thought that I was a salesman trying to sell him rice, and in reply to my question said that his company never bought any Carolina rice. Telling him I grew Carolina rice, I proceeded to consult him regarding my idea of putting it in packages. Very frankly he told me that he hoped rice planters would not do this, for it would interfere with the business of such houses as his, which bought rice in one hundred pound sacks and later put it into paper packages. This of course interested me, and I asked if he would kindly show me one of these packages. For a moment he hesitated, and then reached slowly up on a shelf and handed me a package. I glanced at it and saw that it had "Carolina Rice" written in large letters on all four sides. Upon leaving, I thanked him for the interview and remarked that I was not surprised his firm did not buy Carolina rice, for I saw that they had been able to manufacture it in New York far cheaper than I could grow it in South Carolina.

The principal reason that rice has not been consumed more generally throughout this country is that so few people know how to cook it. Every old Negro woman on Combahee can cook rice better than the best chef in a fashionable New York hotel. When she finishes steaming the rice, every grain stands out by itself, and also the peculiar flavor of Carolina rice speaks for itself.

In this connection, I have never forgotten what I was told by a friend, who for years had planted rice in South Carolina and had later moved to Texas, spending much of his time in the rice section of that state. He said that on one occasion he had attended a banquet given by the Rice Association of

America. Every member present was a rice planter, and yet there was not a dish of rice on the table!

I doubt seriously whether the planters in Louisiana, Texas, and Arkansas are yet eating, to any extent, the rice they grow, while those of us in South Carolina and Georgia, who know what good rice is, are now compelled to eat their rice, or none at all. Even an old Combahee Negro who has cooked rice all of her life cannot make their rice taste like our old Carolina rice in years gone by.

It was on account of these combined causes I have mentioned: namely, inefficient labor, over-production of rice in the Southwest, changes in methods of marketing, and the frequency of storms, which gradually forced our planters to give up the hopeless struggle.

How rice planters did hate to have to give up the planting of rice, especially those who clung to it to the very last! It was their life.

Mr. Heyward Lynah was one of the very last to plant. His home during the latter part of his life was in Savannah, Georgia, but his plantation lay in South Carolina, just across the Savannah River. He visited it nearly every day, and toward the latter part of his life he planted only thirty acres; just enough for him to think of the old days.

Each year at harvest time his family noticed that he hung on the wall of his house on the plantation a sheaf of rice, and removed the one he had placed there the year before. Two years ago, when he died at a ripe old age, he requested his family to place the sheaf on his casket instead of flowers. He wanted the sheaf of rice to go with him to the grave.

25

A WEST INDIAN HURRICANE

THE late Judge H. A. M. Smith, of Charleston, who knew more about the Low Country of South Carolina, its past and present, than anyone else who ever lived in it, several years ago said to me, "A rice planter gambles with Nature, and Nature throws the dice. All the planter can do is to stand and take it." The hazards of rice planting which the Judge principally had reference to were the storms which once so frequently devastated our coastal country and contributed in a very great degree to the abandonment of rice culture in South Carolina and Georgia.

According to accurate records Charleston, in the center of our former rice industry, has for the past two hundred and thirty years been visited by tropical storms on an average of every twelve years. During the twenty-five years I planted rice, from 1888 to 1913, we had a storm every six years, exceeding this record of two hundred and thirty years by 100 per cent. Some years our storms came at comparatively short intervals. There was a time in the past, however, which beat my record. In three successive years in 1850, 1851, and 1852 there was a storm every year. They skipped 1853, but in 1854 there occurred one of the most disastrous storms that Charleston and the rice fields of that section have ever experienced. Four storms in five years. But to offset this, a century earlier there was a period of forty-five years without a storm, from 1752 to 1797. I wonder if the people of Charleston in those years thought the storms had altogether ceased. Perhaps they wished one could have come during the years of the Revolution, when the British fleet lay in the offing.

I read some time ago an old West Indian adage concerning hurricanes, which ran as follows:

> June, too soon,
> July, stand by,
> August, come it must,
> September, remember,
> October, all over.

I have checked this old adage with the weather records of Charleston, and with the exception of the month of October I have found it correct. In the month of June we have never had a storm. In July, only one. In August, nine, usually the worst. In September, sixteen. The adage missed it only in the month of October, for the records show that we have had three storms during that month. Maybe it took our October storms a long time to reach our coast, and the old adage was meant to apply only to the West Indies, where the storms originated.

It is said in Charleston that from its beautiful Battery on a still night a week or ten days before the approach of a West Indian hurricane, the roar of the surf on the sea-islands several miles away can be distinctly heard, although the weather be clear and the stars shining. Should the moon at the time be nearly full and its perigee approaching when the sound of the surf is heard, the residents of the city know that a storm is on its way and with true philosophy await its approach.

Several days before the storm, the wind sets in from the north or northeast, blowing gently at first, and small fleecy white clouds begin to appear. Gradually the sky becomes overcast, though no large black clouds are to be seen; there is only a leaden sky. The force of the wind increases each hour, coming at first in slight gusts while the barometer steadily falls. As the wind continues to stiffen, the tide begins to come higher and to fall very little. White-caps appear in the harbor and the eastern horizon darkens.

All small pleasure craft seek shelter up the Ashley River, and crews of the larger vessels riding at anchor in the bay begin to batten down their hatches, making fast their anchors, and, with everything snug, hope to ride out the storm. The wharves of Charleston present a busy sight. The captains of tugboats and coastwise schooners and ocean tramps all look to their moorings, and take every precaution to keep their ships off the wharves.

Racing into the harbor come the little fishing boats, or "mosquito fleet," manned by Negro fishermen, sailing in, some twenty or thirty of them, their white sails full set and billowing in the freshening breeze. They sail with the wind full astern. Their masts bending and their boats keeling over on their sides, they ride the waves which begin to be dark and white-crested.

As the center of the storm progresses up the coast, the

gray sky continues to darken and the wind comes in stronger gusts from the north-northeast to northeast, and from this quarter it works due east, and each hour blows harder. The wind seems to come out of the bank of heavy black clouds hanging low on the horizon, which gradually spreads, as out of it the rain begins to descend, light at first but increasing as each gust drives it inland.

The velocity of the wind soon becomes tremendous: from forty to fifty, from fifty to seventy-five, and sometimes for a short while reaching ninety miles an hour, while a blinding rain comes down in great slanting sheets, striking with almost irresistible force.

Charleston always received the full force of the wind, for nothing stands between it and the angry ocean. Leaves blow before the gale; limbs are broken and trees blown down; business signs are carried away; and roofs are stripped off houses and many small buildings crash down. Through the entrance of the harbor the ocean rushes as the storm gathers up the water of the Gulf Stream and sweeps it toward the coast. Waves pound against the sea-wall of the Battery, their spray rising high in the air and being driven the distance of a city block. All the while the ocean keeps coming in and, as the wind from the east increases, ships drag their anchors and are driven ashore, while many are pounded against the wharves. Except for the dim light of the hidden moon, the city is in darkness.

When the fury of the storm reaches its height, suddenly, when it seems impossible for anything to stand longer before it, the wind dies down, and for a short time, except for the rain, everything is still. The wind soon comes back again, but from the south, for during the lull it has shifted. The air seems warmer and the waves stop pounding so hard, though they still keep rolling in. Soon the rain ceases and the next day the sun shines hot and scalding through the damp air. Many citizens of Charleston come out on the streets sightseeing, especially along the waterfront. They note the marks which the water reached and compare them with the marks of former hurricanes, for Charleston has experienced many a blow and its citizens are not easily disheartened or discouraged. Storms, sieges, fires, and earthquakes have done their worst, but the old city has withstood them all. The chimes of St. Michaels have never ceased to ring, until very recently; from its belfry at midnight the watchman has never failed to call "all is well."

For days succeeding a West Indian hurricane a scene of desolation prevailed on our rice plantations. Heavy-hearted planter and overseer rode its river banks where riding was almost impossible. They forced their horses through tangled vines and briars, which had been blown across the bank, the horses scarcely able to keep their footing. Outside of the bank, the river is slowly falling, for the wind is from the west, but its water, through the open breaks in the banks, still flows into the fields.

Within the river bank, where only a short time before an inspiring sight had met the eye, a sorry spectacle is presented. Where the crop had been cut and stacked, the stacks have sunk in and stand deep in water, sprouting and rotting under the September sun. Nothing can be done until the breaks are mended.

The planter stops his horse and gazes across his fields,

hidden beneath the water. Past years come back to him, and his eyes have a far-away look as he wonders what those who once planted these fields would do if they were in his place. Would they mend their breaks and plant again? He feels that they would, and so must he.

With the west wind blowing and the sun shining, he recalls an afternoon years ago, when first he rode those banks and saw a rainbow in the sky. In his imagination he sees it again, its end resting in his fields, and hope fills his heart. Then and there he determines to mend his breaks and plant again.

I know from my own experience during the last years of rice planting in South Carolina that a planter whose crop had recently been destroyed by a West Indian hurricane dreaded having to go to Charleston for the purpose of borrowing money with which to start another crop, far more than he had minded the storm itself. There was always a certain amount of excitement while the storm lasted, so much room for speculation as to the damage it would do. How greatly would the velocity of the east wind increase? How much higher would the tide rise? How many breaks would there be in his river banks? How soon would the wind veer around to the west? But, after the storm had done its worst, having to go to Charleston, with his business and everything he possessed depending upon the result of his visit, seemed a far more serious matter. Would he be able to plant another year? Had the end finally come? His trip would decide it. I can recall most vividly the scene.

A cold morning in late autumn. The sun, like a great ball of fire, appearing through the mist, can be dimly seen as it rises over the fields where the warmth of its rays will soon begin to dissipate the white frost on the dead grass and the rotting rice. No bells ring that morning to call the hands to work. Idle for several months, the Negroes loaf all day in the settlement. A few of them, in the cleanly swept front yards of their houses, beat with mortar and pestle the little damaged rice they have managed to save. The mules, reduced to half feed, forget there is such a thing as a foreman's whistle, and as idle as their plowmen in the settlement street, they stand in the stable lot warming themselves in the early sunlight.

After an early breakfast, eaten in silence, wife and children bid the planter goodbye and wish him Godspeed on his mission. Faithful old Joe has his buggy waiting at the front door, but that morning Joe is not allowed to drive. He doesn't understand the importance of catching the train, thinks "crap bleeged fuh plant," might miss the train.

Through the avenue, beneath the limbs of the live oaks. Great old trees, covered with gray moss, planted by his great-grandfather long ago. One backward look at his home, the home-site of his family for generations, granted by authority of the Crown. What is to become of it and of the fields his forebears reclaimed? His heart begins to fail him for a moment when he reaches the head of the avenue and drives between the two brick gateposts standing there. When a boy, he used to swing on the old gate just as his children swing now.

The polite salutations of the few Negroes he meets on the road or finds loitering on the depot platform cheer him up a little. To them he is still the "Boss." Still has plenty of money. "Wadduh gone wid 'e grampa money? Boss mus be hab um." He knows these Negroes and calls them by name. A common past binds him and them together. If he cannot continue to plant, what will become of them as well as of him? Where will they go? What will they do?

Seated in the smoking compartment of the day coach, he finds two wholesale grocery salesmen. Complain of hard times. No business. Texas rice three cents per pound. No Carolina rice to be had. Gazing stolidly out of the car window, he becomes even more dejected. Knows he hasn't a friend in the world.

Arriving in Charleston, he takes a street car down East Bay. His fellow passengers, to him, look so prosperous and happy. Talk so cheerfully. He passes drays loaded with cotton bales, bumping over the old cobble stones. Wishes he had planted cotton. And then his factor's office, and the dreaded interview. Result: a lien is signed, another mortgage to be recorded. He will plant another crop! Fifty bushels to the acre!

How brightly to him the sun now seems to shine, and what a weight is lifted from his mind! His spirits, how they rise! With a glad hand he greets his friends along East Bay and on Broad Street. Such pleasant, genial people, they know all about him. All about rice planting. So much in common. And then to the Club. Such congeniality. Such gentlemen. So appreciative of a gentleman. Delightful change. Delightful day.

That afternoon, on his journey home, the planter is a different man and looks out on a different world. On his way to the depot, the drays hauling cotton annoy him. Poor way to handle a crop. Schooners freighting rice to Charleston discharge alongside the mills. At the depot he enters the Pullman. Talks to some Northern tourists, so interested in Charleston and the country the train passes through. The names of the stations they pass and of the rivers they cross—Indian names, nearly all—excite their curiosity. Fine rice plantations on some of these rivers. Safe plantations, he tells them. Forty bushels to the acre, sixty bushels, world-renowned Carolina rice. Storms? Happen now and then. Record of plantations. Storms come in cycles of years. End of a cycle. Better day dawning for rice planting. More appreciation of Carolina rice. Thank God for Captain Thurber and the seed from Madagascar!

And as he steps off the train at his little station, the planter, like an English king of old, is himself again. Seeing some of his Negroes on the platform, he calls to them, "Boys, want you all out tomorrow. Begin work. Send out the word." Old Joe in his buggy patiently waiting. Little now does he care for the holes in the causeway. Up the old avenue of live oaks. Light from the opening door. Within, the cheerful pinewood fire. Welcoming faces, anxiously waiting to know the result of his visit.

And that night as darkness comes to his home, more intense on account of the live oaks and magnolias which surround it, and a chilly dampness, deepening with the coming of the stars, settles along the highland, and a white mist hides the waiting fields, hope again reigns in the heart of the planter. In his dreams he hears the singing of a mocking-bird in the magnolia near his window; on horseback he rides his checkbanks between squares of green, waving rice, which day by day from his saddle he watches as they turn into fields of golden grain. In the distance, the sails of a schooner, headed up the river, bound for his threshing mill where a group of Negro children wait with their "fanner baskets" to fill the hatches to the brim. With the rising of the morrow's sun, the dreamer will awaken to make ready for the sowing of the seed. Once more will the planter put his trust in the Lord of the Harvest.

26

MENDING BREAKS

WHEN a storm was threatening, a rice planter, riding through his fields, was always apprehensive when he saw a Negro on a horse galloping in his direction. His heart seemed to stop beating, for he immediately thought of a break and sometimes he thought aright. "Boss, we hab a berry bad happedence. De bank in seed square broke. Me ben dey en schum mesef."

Breaks in the river banks were usually caused by high tides. The water as it poured over the bank cut it away until it became too weak to withstand the pressure of the river. Even without running over, a very high "spring" tide might cause a bank to push in at some weak spot, and the water rushing in soon widened the gap. Most of the disastrous breaks, however, were caused by storm tides.

The banks protecting the fields were usually not more than two feet above normal high tide, and it may seem strange that the planters, and especially those in the days of slavery, did not raise their banks to a greater height. The reason is obvious. The height of a bank cannot be safely increased without also increasing its width, thus giving the bank its proper slopes. This could not be done by those who planted after the Civil War, for it would have necessitated the use of expensive dredges, which they could not afford. As to the men who first built the banks with slave labor, the fact must be remembered that when the rice lands were reclaimed, they were considerably higher than they were in later years, and neither a higher nor a heavier bank was so necessary. Furthermore, all of the work on rice plantations in those days was done by hand, and the Negroes were kept busy all the year in cultivating, harvesting, and threshing the crops, and little time was left for bank work. Besides, a break in a bank was by no means as disastrous in the early days of the industry as it became later, for with their great supply of labor fully under control, the planters could repair a break very quickly, unless it was an unusually large one, before much damage was done. Keeping their banks in good order was all they deemed necessary.

There is nothing that gives one a more helpless feeling than to stand at high tide on the crumbling edge of a break which has recently occurred, and to watch the water sweeping through the crevasse it has made. The break may be only a small one, from twenty-five to fifty feet, but nevertheless the water looks terrifying as the tide, with the full pressure of the river and ocean behind it, on reaching the opening turns from its course and surges down into the field with great force.

Should the break take place where there is a wide marsh, the water flowing over this soon cuts it away and carries the marsh into the field in large lumps, scattering them in every direction. Nothing then stands between the field and the river, and until the tides begin to decrease, field after field is flooded, until the whole plantation is submerged. A little later on, when the tides go lower, the water runs out of the field twice a day, for a short time, during which it cuts a large hole in front of the break. Very soon, both this hole and the break are as deep as the bottom of the river, for both day and night the water is either coming in or going out of the field.

Frequently after a severe storm a rice planter finds that in addition to the total loss of his crop he has several breaks on his plantation, and to save his property from complete destruction, these breaks must soon be mended. He fully realizes, too, that looking at the breaks, talking with his neighbors about them, and seeing them in his sleep are futile. Only by going to work on them can anything be accomplished. Help can come from no source, the work cannot be let by contract, for contractors are usually afraid to undertake such work. The planter alone understands what will have to be done, and he only can manage the labor who will undertake the job.

He must first decide how the break had best be stopped, for there are two ways, either by what was termed a "straight stoppage" or by building a "half-moon." A straight stoppage necessitated the use of very large piling, and consequently a heavy pile driver, for it meant rebuilding the bank where it stood before, and where the water had become quite deep. During the last twenty-five years of the industry, this method was generally abandoned, for it required a long time to complete the bank and, in addition, was exceedingly expensive. Nor was it at all certain to be successful, for unless the piles were deeply driven, which was difficult to do on account of sand in the bottom of the break and the rush of water, the whole structure was liable to capsize section by section, even after it was nearly filled with earth. Resort, therefore, was had to what were known as half-moons, as the quickest, safest, and cheapest method, especially when the lay of the field was such that when the tides became normal, the field went dry for a short time.

The half-moon seemed a very simple structure, but both experience and judgment were required to construct it, for one had, except for a short time each day, to contend with a strong movement of water in and out of the field, and a planter had to know just what to do, as well as what not to do, for if a mistake was made, much of his labor and material would be thrown away.

No actual work can commence until posts and lumber are accumulated and until the tides subside, for the dirt with which a half-moon is filled must come from the field. When the proper time comes, all the able-bodied men on the plantation are called out, and others come from adjoining plantations and from their little farms in the pineland. Often they have to be carried to work on flats on the river, the checkbanks in the fields being under water, for if there is one thing a Negro dislikes, it is getting wet early in the morning while going to his work. In stopping a break, however, few object to getting wet through and through, when the work is on and the excitement runs high.

As soon as all arrive on the spot, the posts are quickly sharpened and, beginning near each end of the break, about ten feet from its edge, are placed in position. They are set in line five or six feet apart, and another line of posts is set ten or twelve feet from them. The posts are immediately opposite each other. Quickly these two lines of posts are capped across and fastened together with two-inch planks, firmly spiked. The posts are placed in a semicircle, leaving the deep excavation on the outside and being kept well away from it, so much so that the half-moon is usually five times as long as the width of the break. As soon as the caps are on, planks are laid across them, and the process of driving down the posts begun. For this purpose a section of some hardwood tree is used, into which are fixed handles made from a stout grape-vine. The Negroes called this a "baby" and it requires four good men to lift it. One man gives the "cry," at which all lift the weight as high as they can and let it fall on the head of the post, usually driving it down from six inches to a foot each blow, thus making the framework steady and fairly strong.

As soon as the first posts on each end of the half moon are capped and driven down, two-inch planks are nailed to the inside of the posts, forming a box, as it were. Then runways are made, leading up from the field to the frame work, and immediately the wheelbarrows are started. Lines of Negroes, straining between their handles, come up the runways, and one after another dump their barrows into the stoppage. Each hand is paid by the barrow, and each is given a number which he calls as he dumps his barrow, so that a correct record may be kept. Often a prize is given to the Negro who rolls the largest number of barrows, and thus an almost constant stream of dirt is thrown in. Quick work counts, for shortly after the tide turns, the hands are driven from the field and nothing more can be done until the next day. In this way, each day one or more sections are filled, and it must always be as long as the width of the break itself and be left in the center of the half-moon. It is also necessary to finish this section in one day, for should the incoming tide be restricted as it rushes through it, all that had been already built would be washed away.

For the last day, every preparation is made. Large iron pots are brought and dinner for all is cooked on the bank, consisting usually of rice, peas, and bacon. Often a demijohn of corn whiskey is placed on the bank in plain sight, but carefully guarded, and all hands are promised a drink when the water is shut out, but never before. Long before the tide leaves the field, a foreman, with a few picked hands, jumps into the water and begins getting the plank sides ready, for the time is short and there can be no delay. Every Negro realizes he has to get wet.

As the land goes dry and the barrow-men begin rolling in mud, barefooted and with their pants rolled up to their knees, wet and muddy, they follow each other up the runways on which half hands sprinkle sand, for rice field mud is as slippery as a banana peel, and it is also so sticky that one walking about on a half-moon while it is being built soon accumulates enough mud on his boots to make a small farm.

When the tide turns and begins to rise, a race is started between the barrows and the water. The Negroes are urged on and encouraged, and it must be said to their credit that they usually do their very best. In the emergency, they always respond heartily, joking with each other, as they empty their barrows and return for another load, straining every nerve to keep ahead of the rising water. At last it becomes evident that the wheelbarrows have won, as the dirt rises higher and higher, and its base becomes broader. Then the foreman gives a signal and the barrows stop. A dram is distributed in a tin cup and the dinner follows. Within a few days, the new bank is raised again and rounded out, and for years it stands, mutely telling its tale to those who can understand. Few half-moons ever give way.

In mending a break the main consideration is always where to obtain dirt, for in a rice field only dirt can check the flow of water. Where it has been found impossible to dig in the field near the break, I have known earth to be brought in flats for miles or on pontoon bridges made of flats strung across the river, over which earth was brought in wheelbarrows.

In my experience in rice planting, Mr. Jaycocks and I failed to stop but one break, and that I have always maintained was not our fault, for we had closed all but the middle section when a most unlooked-for occurrence suddenly stopped all work. The low tide came in quite late that afternoon, and it was decided to finish the break that night, for the reason that the moon was full and there was not a cloud to be seen anywhere. The Negroes all agreed to work by moonlight and supper was prepared. The number of boys sprinkling sand on the runways was doubled, and every precaution was taken. The moon rose bright and clear, and by its light the hands had no difficulty in working. The barrows came rapidly, and the thud of the falling dirt, as it was dumped, seemed louder on the night air. Soon we were well ahead of the rising tide, and everything seemed safe, when suddenly it began to grow dark and the barrows stopped. "Great Gawd!" I heard one Negro exclaim, "Wat is dis?" as he slipped and with the wheelbarrow splashed headlong into the water. Looking up, I saw that the moon was rapidly going into a total eclipse! If

I had seen this predicted, I had forgotten it. Soon the tide swept over the unfinished section and in a short time frame and all were washed away. On the next day, all had to be begun over again.

Going home on the river on a flat late that night, wet, cold, and tired, the muddy Negroes lying fast asleep around me, it did seem that the Almighty Himself had not only forgotten the rice planters but was determined to wipe them out.

27

OUR LAST TWO DESTRUCTIVE
STORMS

I FIRST learned of equinoctial storms from Mr. Jones, who had spent his entire life on, or near, the Combahee. And from him, when I began planting rice, I got the impression that scarcely more than once in each generation had my family greatly suffered from these storms.

From the year 1855 until his death in 1897, Mr. Jones kept in close touch with my plantations. As to storms, he knew the records of these plantations, and also those of Rose Hill and Pleasant Hill, Charles Heyward's other plantations on the Combahee. He often told me that from 1854 to 1893 there had been no year when these plantations had failed to harvest enough rice to pay expenses. During that time, a few storms had just missed Combahee and several had damaged Charleston. For thirty-eight years Combahee had been free from storms that did any great damage, but in the years which followed, the story was very different.

Sunday and Monday, August 27 and 28, 1893! If for thirty-eight years prior to 1893 no very destructive storms had visited South Carolina, those days alone made up for it, especially on the Combahee. The very center of this storm struck Beaufort, and on the sea-island adjacent to Beaufort the loss of life was greatest. Charleston was very badly damaged and a few of its citizens were killed, or drowned. People in Charleston today who can recall this storm say it was one of the worst Charleston ever experienced, and I can well believe it. Rice, and all other crops, between Charleston and Savannah were completely destroyed. For the first time, the services of the Red Cross, then only recently organized, were asked for in our state. And I recall seeing Miss Barton on the train one morning going to Beaufort. For two or three months after the storm, the Negroes from Combahee flocked to Beaufort to lay their claims before the Society. I remember one day hearing some Negro women laughing at a man, who was not by any means noted for his industry. I asked why they were laughing at him and one replied: "Great Lord, Boss! You no John tak a foot and wak all de way to Beefut fur see day Red Cross ledy wha gib way bitual and ting, an all de buckruh 'ooman givum bin a hoe. John lef um weh he git um."

I was in Lexington, Virginia, when this storm occurred and reached Charleston two days later. The train from Combahee could not get further than Ashley Junction, seven miles from the city, for portions of the track of the railroad to Savannah were still under water, which gives an idea how the whole country was flooded. Early the next morning another start was made, and I reached my station.

I remember distinctly stepping off the train and looking around for my horse and buggy. Finally I saw one of my water-minders walking to the station and asked him where the buggy was. "Buggy, Sur?" he said, "buggy! I fetch de boat fur you." A short distance from the depot, we took the boat, and rowed across three plantations. In the fields, the water was so deep that only in certain places could I see the fully matured rice, which was standing upright under water, having righted itself after being blown down. On reaching my plantation, I found a number of breaks in the banks, one being the largest I have ever known on the Combahee. In fact, the plantation was all but destroyed, and it was months before anyone could get around the fields in anything but a rowboat. I did not make a pint of rice, nor did scarcely anyone on the river. There were a number of Negroes on the rice plantations in those days, and it was a great problem for the planters to care for and feed them.

For the four years succeeding 1893 we had good seasons, no storms occurred, and good crops were made. In fact, our rice planters were just getting things in shape again, when on September 29, 1898, we had another storm. This was not a very bad one, but resulted in a break which caused the loss of more than half my crop, which was stacked in the fields.

The most peculiar storm which ever visited the coast of South Carolina was the one in 1910, which came in October. In fact, though the planters have always thought and spoken of the calamity which befell the rice crops on the Combahee and the Edisto rivers that year as a "storm," in reality it was not, for the wind never reached the coast with any great velocity; yet the tide rose as high as I have ever known it to do, with the exception of the storm in 1893.[1] The storm was predicted by the weather bureau, but turned in its northward course, just missing our coast, and went out in the Atlantic,

[1] The Combahee and the Edisto rivers in 1910 were the only rivers in South Carolina on which rice was planted to any extent.

but it came near enough to pile up great tides south of Charleston.

We had a very rainy summer and fall and the swamps on the upper part of the Combahee were full to overflowing. The prevailing wind for more than a month had been from the east, backing much water up the river and allowing little to go out. Full moon and perigee came on the same day. These conditions within themselves would have caused very high tides, but happening in conjunction with the storm in the Gulf Stream, they caused a tremendous one.

The telling of my experience the day of this disaster will give some idea of how a rice crop can be destroyed and will also show one of the reasons why rice is no longer planted in South Carolina.

I had eighteen hundred acres planted that year and made a good crop. I was planting three plantations, each separated from the other by water—the Combahee River and Cuckols Creek. The rice was all cut, but was still stacked in the fields, the harvest being unusually late on account of a rainy fall. Though it was about the middle of October, I had threshed only two days, and on the third day I left Charleston early in the morning and reached my Amsterdam plantation about eight o'clock. The storm signals were up when I left, and the morning paper said a dangerous storm was approaching. The wind was northeasterly, but not blowing at all hard, and when I reached the plantation the sun was shining, though there were some white clouds to be seen.

I can distinctly recall standing in the door of my threshing mill on Amsterdam plantation, and looking over the fields toward Bonnie Hall, across the Combahee River, a large plantation which I had leased. I could see the white steam puffing from the exhaust of the mill on Bonnie Hall, and I knew I could count on a thousand bushels of rice being threshed there that day. To my left I saw the mill on Myrtle Grove plantation, across Cuckols Creek. Its exhaust, too, showed that it was hard at work, and I could count on another eight hundred bushels there. I knew that my old mill at Amsterdam would do even better than the others; its long "conveyor house" was filled with bundles of rice and in it men and women were cutting bands and laying the sheaves on tables from which they were being placed on the conveyor. Down the road leading through the fields was a long line of carts and wagons—little one-horse affairs owned by the Negroes who drove them—filled with sheaves, the drivers sitting on top of the rice. It was, indeed, a busy scene, and one which a planter worked a year to see. Negroes were rushing here, there, and everywhere, excited by the noise of the mill, and the realization that it was harvest time.

In the mill, I found Mr. Jaycocks, and told him the storm signals were up. He looked out of a window and said he did not believe that we would have a storm. Our attention was soon attracted to the water in Cuckols Creek, which was nearly up to the top of the river bank; and the tide, though it should have been high water, was still sweeping up the creek. I immediately took a boat and crossed over to Myrtle Grove, and walked to the mill. Before I reached it, the water was running steadily over the river bank, getting deeper and deeper, though at that time the tide should have been on the ebb for more than an hour. I ordered the mill to be shut down, for the barnyard was covered with water.

I then sent a Negro on his horse to ride a portion of the river bank and let me know how matters stood. In a short time, to my dismay, I saw him galloping back. Getting a horse I went to meet him and he told me there was a bad break in a certain square. I rode through the square, as near to the river as my horse would go, with the water nearly up to his belly. About forty feet of the river bank had gone, and as the water, sweeping up the river with the force of the ocean behind it, came to the break, it rushed down with a roar into the fields, turning over rice stacks as it reached them. I knew the crop on that part of Myrtle Grove was gone, and riding back to the mill, I learned that other breaks had occurred. The whole crop would be a total loss.

I immediately recrossed Cuckols Creek to my old rice mill, which was still running, but the wagons had stopped. There I was handed a note from Mr. John C. Porcher, a most excellent planter and the manager of Bonnie Hall, saying there were three bad breaks there and asking me to come right over. I hurried to the landing on the Combahee, splashing through water and mud, and crossed the angry river in a "trus'-me-Gawd." As I stepped on the Bonnie Hall bank, I saw that the fields were entirely under water, the tops of the stacks of rice barely showing.

Again I committed myself to the "trus'-me-Gawd," and landed safely on the Colleton side. I looked across the fields to the Amsterdam mill. No steam puffed from its exhaust. The mill had shut down. Soon I was joined by Mr. Jaycocks, and we tried to make the round of the plantation. The water-minders had reported no breaks, but there was too much water in the fields to be accounted for in any other way.

Across one of the causeways through the rice field, we found the water sweeping, from a foot to eighteen inches deep. This could only come from a bad break, and that had to be found. Horses now were useless. They bogged to their belly-bands. We turned them loose and tried it afoot, following a check-bank leading in the direction from which the water was flowing. Soon the water caught us nearly to our waists. The sun was intensely hot, and the weeds along the seldom-used check-bank were high and leaning across the bank so that they made walking most difficult. After pushing our way through water and weeds for a mile, we came to the break—a very large one—the water pouring through it. Soon a water-minder reported another break, and by one o'clock that day I knew that the crops on all three plantations were lost, and that there was nothing left for us to do but begin figuring on stopping the breaks. In three hours a scene of activity and prosperity was changed into one of stillness and desolation. Nothing left to the rice planter but another year's hopes.

I planted the same amount of rice on the three plantations again the next year, 1911, and on August 27, ten months from the date of the loss I have just described, a storm struck Combahee, then the only river in South Carolina where rice was grown, for the Edisto planters had given up after their losses in 1910. This storm struck our coast with very great force. I shall not undertake to describe it. All night I sat on the piazza of my house in Charleston, three doors from the Battery, while the wind came in gusts, then came again in greater gusts, driving the rain before it through the blackness of the night.

I saw the ocean actually coming up Meeting Street, until

all the lower part of Charleston was under water. As the water receded, I could see timbers from a wrecked vessel floating down Meeting Street, going out with the tide to the sea. When daylight came the waves still pounded against the high wall of the Battery, striking with such force that their spray was dashed as far as Church Street, a block away. I knew that night that on Combahee we were having a repetition of the destruction of the year before, and I knew also that the death-knell of rice planting in South Carolina was sounded. Two disasters in quick succession—only long enough apart to allow a crop to be grown and ready for harvesting, and then destroyed! Since then, there has been little rice to destroy.

And so, year after year, for the reasons I have given, the planting of rice gradually decreased in South Carolina and Georgia, until by 1914 the growing of rice had practically ceased in these states. On rivers whose plantations had shipped thousands of bushels to market, scarcely a single threshing mill was in operation, and the only evidence we see today of where these mills stood is their brick chimneys. Charleston and Savannah are no longer rice markets, except for rice grown beyond the borders of their respective states, and their pounding-mills are closed for all time. Schooners, loaded with rice for these ports, have vanished from every river which flows into the South Atlantic.

After the storm of 1911, I was compelled to give up planting Amsterdam and Lewisburg plantations, and also Rotterdam, which I had recently purchased. Two plantations owned by relations of mine, one on the Bluff, the old home of Nathaniel Heyward, and the other, Hamburg, then owned by his great-grandsons, and both protected by the same river banks as were my plantations, were not to be planted the next year. The length of the protecting bank was seven miles, and a break in it on any one plantation would seriously affect the others. I could not arrange to plant my places under these circumstances; the risk was too great and I was forced to let them lie idle. The next years I planted a small acreage on Bonnie Hall and Myrtle Grove.

As often as I have been asked why rice planting was abandoned in South Carolina and Georgia, I have been asked another question: "If our planters failed planting rice, why did they not plant other crops on their fertile lands, lands capable of being so readily irrigated?" With the exception of indigo, which was planted for a short time only, no other crop has been successfully grown on our former rice lands under the conditions in which rice was grown. A prominent planter on the Cooper River used to claim that one could lose less money planting rice in a rice field than he could lose planting anything else, and he was quite right. The planters in those last days of the industry had no choice. Not being able to obtain advances with which to plant, they were forced to let their plantations lie idle.

Very shortly after this, I sold my three plantations to Northern capitalists. The Bluff and Hamburg were also included in the sale. A company was formed to promote the cultivation of other crops on our former rice fields. To do this with safety, it was necessary to increase greatly the size of the river bank protecting the entire property; to dig large canals through the fields; and to drain them by a large pumping plant centrally located. This was an undertaking I had thought of for a number of years, the feasibility of which had been approved by the United States Department of Agriculture. But this is another story and would require too long in the telling.

In interesting these Northern capitalists, I had three objects in view. I believed the proposition of planting other crops on our rice fields was fundamentally sound and, if successful, would be of untold benefit to the Low Country of South Carolina, where my heart has always been, and I wanted to see it come back into its own. Especially did it hurt me to see abandoned the lands which had been reclaimed by my people and planted by them for so many years. I thought also of the Negroes on those old plantations for generations. They had been faithful to my people and they had been faithful to me. Unless something could be done which would give them work, they would be faced with starvation.

28

"ME MAUSSUH 'E ENT HAB NO LAN'"

ONE day after the transfer of my property had been made, I happened to be on Combahee and was sitting by myself on the piazza of the dwelling at the Bluff when old Judy Simmons chanced to pass by, saw me, and stopped to talk. We talked of old times, of my father and my grandfather, and finally she asked what I expected to plant that year, and I told her, nothing; that I did not own any land to plant. Immediately the old woman began to weep and exclaim, "Maussuh, don' tell me onnuh ent hab no lan'." I assured her that it was true. And, finally, still weeping and wringing her hands, she turned away and walked slowly down the long avenue of live oaks leading from the house. And as far as I could see her she was still weeping, waving her arms, and crying at the top of her voice, "Oh, me Gawd! Me Maussuh 'e ent hab no lan'. 'E ent hab no lan.' " Repeating it again and again.

I sat there on the piazza watching her until she reached the end of the avenue where the old oaks seemed to draw nearer together, their streamers of gray moss reached closer to the ground, and the figure of Judy grew smaller and smaller. I watched her until she faded from my view, and as I did so, I thought of Ike Davis, who had died years before; of our arguments at school about slavery; of our talk that day in the road with old Judy, and how I wished the friend of my youth could hear that lament of an old Gullah Negro, who for many years had been a slave. "Oh, me Gawd! Me Maussuh 'e ent hab no lan'. 'E ent hab no lan'."

I realized that day, as never before, that at last my father's prediction in regard to rice planting in South Carolina and Georgia had come true.

INDEX

INDEX

Adams, Joe, 59
Affy, 55, 135
Alston, Capt. William, Washington's comment on rice plantation, 17, 18
Altamaha River, rice plantations, 5
Amsterdam plantation, 56, 67, 78, 80, 90, 96, 132, 151, 152, 155, 158-59, 202, 205, 242, 243, 244, 256
Anderson, Betsy, 145-47
Ayllon, Velasquez de, exploration mentioned, 101

Barnwell, Emma, marriage, 91
Beaufort, capture of, 128, 129; storm destruction, 239
Bennett's Rice Mill, 24, 69
Blanford plantation, 197
Bluff plantation, 67, 68, 69, 71, 73-74, 90, 99, 246, 247, 249
Boggs, William, 164-65
Bonnie Hall plantation, 242,-43, 247
Boston, head carpenter, 113
Butler, Gen. B. F., scheme for proclamation, 136

Cape Fear River, rice plantations, 5
Charleston, Heyward residences, 63, 99, 100, 107, 153; residence, 68, 69; Negro nurses, 188; power from tides, 23; in Revolutionary War, 64; storm destruction, 221-26, 240, 245
China, influence on rice culture, 8-10
Civil War, capture of Beaufort, 128, 129; Sherman's march, 69, 135, 136, 199; Emancipation Proclamation, 136
Clarissa, child nurse, 103, 135, 195
Clay Hall plantation, 79

Clinch, Catherine Maria, marriage, 148
Clinch, Duncan L., mentioned, 148
Columbia, agreement with Freedmen's Bureau, 138-42; Barnwell Heyward at college, 148; burial of Barnwell Heyward, 158; captured, 35; church founded by Negro, 195-200; incident during author's governorship, 194, 195; Reconstruction conditions, 151
Combahee River, rice plantations, 5, 14, 68, 100, 101, 241, 245; slaves moved from, 131; yield of rice, 42
Cooper River, rice plantations, 5
Copenhagen plantation, 67
Cotton, introduction mentioned, 48

Davis, Isaac R., argument over slavery, 160, 161, 249-50
Daws Island, landing of French, 52
Driver, slave, duties, 102, 157, 158, 179, 180
Du Pont, Felix A., mentioned as owner of plantation, 25, 162

Edisto River, rice plantations, 5, 241, 245

Factors, business with planters, 85, 86, 217, 218
Flail, use of, 21, 184
Ford, Henry, rice mills, 24, 25
Fraser, James H., mentioned, 43

Gadsden, Lucius, Negro engineer, 25-26, 71, 205
Gibbes family, 66
Gignilliat, Elizabeth, mentioned, 62
Goff, Caesar, carpenter, 113

· 253 ·

INDEX

INDEX

Theus, painter, 45
Thurber, Capt. John, first rice seed, 4, 7
Trapier, Gen. J. H., mentioned, 151
Tucker, John Randolph, tariff on rice, 43, 44

Waccamaw River, rice plantations, 5, 17
Washington, George, comment on rice plantation, 17, 18; visit to Thomas Heyward, 51, 52

Wasp, the, 157, 204
West Point Mills, 24
White Hall, 51-52, 173
Winnowing, methods, 21
Woodward, Dr. Henry, first rice seed, 4, 5; mentioned, 91
Wright, Sammy, 164
Wright, Titus, 172-75

Yellow fever, mentioned, 92